Democratizing Application Development with Betty Blocks

Build powerful applications that impact business immediately with no-code app development

Reinier van Altena

BIRMINGHAM—MUMBAI

Democratizing Application Development with Betty Blocks

Group Product Manager: Alok Dhuri
Publishing Product Manager: Harshal Gundetty
Senior Editor: Kinnari Chohan
Technical Editor: Pradeep Sahu
Copy Editor: Safis Editing
Project Coordinator: Prajakta Naik
Proofreader: Safis Editing
Indexer: Tejal Soni
Production Designer: Ponraj Dhandapani
Developer Relations Marketing Executives: Rayyan Khan and Deepak Kumar

First published: February 2023

Production reference: 1270123

Published by Packt Publishing Ltd.
Livery Place
35 Livery Street
Birmingham
B3 2PB, UK.

ISBN 978-1-80323-099-3

www.packtpub.com

Contributors

About the author

Reinier van Altena started about 9 years ago in the no-code/low-code space, as a no-code developer in Betty Blocks. From building applications for 5 years as a no-code developer, he then progressed to become a lead developer on projects that are built on the platform. Then, he was a Betty Blocks trainer for 2 years, traveling around the globe and teaching business users how to be citizen developers and no-coders.

Now, he works on the product itself as a product owner, using his experience to make Betty Blocks a better product for citizen developers and no-coders.

About the reviewers

Erwin Kenter is a cheerful person who always walks into a room with a smile on his face. He started his IT career 15 years ago and was one of the first to start building apps with the Betty Blocks platform. During this period, he built hundreds of Betty Blocks applications and progressed to become a lead developer, leading a team of Betty Blocks developers that would build bigger and better Betty Blocks applications. With his knowledge, he helped the product team to improve the platform and introduced new features to enhance the platform.

In the last few years, he has switched roles to become a technical consultant, where he trains new developers so that they can build complex Betty Blocks applications on their own.

Robin Jan Wouter van der Burght has more than 10 years of experience in frontend development and design for websites and (web) applications. He has worked for different companies that service different multinational organizations. Currently, he works as a technical consultant at Betty Blocks, where he shares his development experience and insights with clients who build applications with the Betty Blocks platform.

Table of Contents

Part 2: First Steps on the Betty Blocks Platform

3

4

5

6

Creating a New Application from a Template   91

7

Prototyping an Application   115

Part 3: Building Your First Application

8

The To-Do Application   137

9

The ToDo Application – Actions and Interactions   155

Part 4: The Pro-Coder

13

The Back Office   249

Preface

Betty Blocks is a platform aimed at citizen developers and no-coders to help them build their own applications without the help of a pro-coder, while pro-coders can build applications at a faster pace.

I've tried to put together all my knowledge to get you started on the platform and to help you understand how you can get the best result using a no-code platform like Betty Blocks.

I hope you enjoy reading the book and it will help you to start building applications that help you in your day-to-day work.

Who this book is for

This book is for everyone who wants to start building applications for their daily work. Betty Blocks is aimed at citizen developers and no-coders, which means that everyone should be able to learn how to use it, even if they are not a developer by trade.

What this book covers

Chapter 1, What You Can Build with Betty Blocks as a Citizen Developer, offers an understanding of who can build on the platform and what kind of applications can be built with the platform.

Chapter 2, Collaboration between Citizen Developers and Coders Using Betty Blocks, explains how citizen developers and coders can collaborate in developing on the platform. It also talks about whether there is any value for pro-coders on a no-code platform, and what they can offer citizen developers.

Chapter 3, Governance on the Platform, explains how governance on the platform allows you to create organizations, applications, and sandboxes. All of this is governed by different user types. This will be explained in this chapter.

Chapter 4, An Introduction to Data Modeling, teaches you what you can do with the Back Office, including how to set it up and how to access your data there.

Chapter 5, An Introduction to the Page Builder, teaches you what you can do with the page builder, including how to create, wireframe, and build your own first page.

Chapter 6, Creating a New Application from a Template, introduces you to creating an application by using a template.

Chapter 7, What Went into Building This Application, shows you how the application from the previous chapter was built.

Chapter 8, The ToDo Application, gives a deeper look into the Betty Blocks platform. Now that you are more familiar with the platform, we'll dive a bit deeper into pages and data models.

Chapter 9, The ToDo Application – Actions and Interactions, shows you how to add actions and interactions to the ToDo application.

Chapter 10, Case Management Application, shows you how to build an application with a remote data source; we'll grab the remote data source and get the data from there instead of from our data model.

Chapter 11, Case Management – Pages and Actions, shows you how to add pages and actions to your case management application.

Chapter 12, The Pro-Coder Features, shows the features that a pro-coder can use.

Chapter 13, The Back Office, explores the Back Office, which is an older part of the platform and is used for managing your data.

> **Important note**
> The official name "no-coder" has now been changed to "Business technologist" by Gartner.

To get the most out of this book

If you are tech-savvy and good at Excel, for example, a platform like Betty Blocks is not hard to learn, but any business user should be able to create an example application, such as a mock-up, with a little training. The following table lists the requirements to get the most out of this book:

Software/hardware covered in the book	Operating system requirements
Betty Blocks	Windows, macOS, or Linux
A modern browser	

Download the color images

We also provide a PDF file that has color images of the screenshots and diagrams used in this book. You can download it here: `https://packt.link/2vItV`.

Conventions used

There are a few text conventions used throughout this book.

Bold: Indicates a new term, an important word, or words that you see onscreen. For instance, words in menus or dialog boxes appear in **bold**. Here is an example: "Here, click on the list item to open the **Data model** overview page."

> **Tips or important notes**
> Appear like this.

Get in touch

Feedback from our readers is always welcome.

General feedback: If you have questions about any aspect of this book, email us at customercare@packtpub.com and mention the book title in the subject of your message.

Errata: Although we have taken every care to ensure the accuracy of our content, mistakes do happen. If you have found a mistake in this book, we would be grateful if you would report this to us. Please visit www.packtpub.com/support/errata and fill in the form.

Any errata related to this book can be found at https://github.com/PacktPublishing/Democratizing-Application-Development-with-Betty-Blocks.

Piracy: If you come across any illegal copies of our works in any form on the internet, we would be grateful if you would provide us with the location address or website name. Please contact us at copyright@packt.com with a link to the material.

If you are interested in becoming an author: If there is a topic that you have expertise in and you are interested in either writing or contributing to a book, please visit authors.packtpub.com.

Share Your Thoughts

If this book is helping you improve your skills, we'd strongly suggest leaving a review on `Amazon.com`. This helps us know if you like our work and if the chapter content has been valued, and also helps the buyers on Amazon know if the book is right for them. Scan the QR code below to go straight to the Amazon review page for this book and share your feedback.

https://packt.link/r/1803230991

Your review is important to us and the tech community and will help us make sure we're delivering excellent quality content.

So, everyone else benefits from your review - we wouldn't want you to miss out. You can now reach out to `review@packt.com` with a screenshot of your review and the book URL, and we'll send you a $5 voucher for your next Packt purchase. Thank you in advance for engaging with us, we are excited to see your review!

Download a free PDF copy of this book

Thanks for purchasing this book!

Do you like to read on the go but are unable to carry your print books everywhere? Is your eBook purchase not compatible with the device of your choice?

Don't worry, now with every Packt book you get a DRM-free PDF version of that book at no cost.

Read anywhere, any place, on any device. Search, copy, and paste code from your favorite technical books directly into your application.

The perks don't stop there, you can get exclusive access to discounts, newsletters, and great free content in your inbox daily

Follow these simple steps to get the benefits:

1. Scan the QR code or visit the link below

https://packt.link/free-ebook/9781803230993

2. Submit your proof of purchase

3. That's it! We'll send your free PDF and other benefits to your email directly

Part 1:
Citizen Development

In this part, you will learn what citizen development is and what citizen developers can do. This part has the following chapters:

- *Chapter 1, What You Can Build with Betty Blocks as a Citizen Developer*
- *Chapter 2, Collaboration between Citizen Developers and Coders Using Betty Blocks*

What Can You Build with Betty Blocks as a Citizen Developer

This chapter is an introduction to Betty Blocks and citizen development. We'll go over what the Betty Blocks platform is and what you can do with it as a citizen developer. We'll also talk about the different types of developers that utilize the platform and what kind of roles they each have in the development of an application. Along with this, we'll examine the kinds of applications you can build with Betty Blocks, so you'll get an idea of what the platform is capable of.

In this chapter, we will cover the following topics:

- Introduction to the Betty Blocks platform
- Who can build applications on the Betty Blocks platform?
- The different developer personas
- The type of applications you can build
- Where do applications on the Betty Blocks platform reside?

By the end of this chapter, you will have a high-level understanding of the platform's functionalities, what the different developer personas can bring to the table when developing applications, and how they can utilize the platform.

Introduction to the Betty Blocks platform

Before we jump into what you can build with the Betty Blocks platform, let's discuss what Betty Blocks is. Betty Blocks is a no-code application development platform hosted in the cloud, so you can access it from anywhere you have internet access.

This means that you can develop applications in your favorite browser from anywhere in the world. Betty Blocks provides you with an all-inclusive tool that allows you to store data, design pages, create workflows, and interact with external services such as APIs, OData, and OpenAPI. All of these form an intuitive user interface that mostly works with drag and drop.

The platform is made for all kinds of developers. Betty Blocks identifies three specific types, namely the *citizen developer*, the *no-coder*, and the *pro-coder*. All of them have something to add to the process of developing an application and need to collaborate with each other to produce the full application. We'll dive deeper into how this collaboration works between the roles and who can do what.

Because the platform is aimed at citizen developers, you might be wondering whether this mean there is no learning curve? Well… of course there is! Think about people using Excel, for example. Anyone can start inputting data into Excel, but to properly use a lot of its functions, some training might be required. The same goes for Betty Blocks: you might be able to do a lot of basic things already, but to understand all the functions that you can add to an application, it helps to go through a course. This is likely why you are reading this book.

Do *no-code* and *drag and drop* mean that you'll be restricted in terms of what you can do with the platform? The answer to that is no. We'll dive deeper into this later, but what it comes down to is that the platform allows experienced developers to add code to the platform as well, which increases the capabilities of the platform and thus also the options citizen developers have within the platform.

Who can build applications on the Betty Blocks platform?

The main goal for Betty Blocks is to enable business users to build their own applications with the platform. But within that group of business users, there are different levels of developers, namely the citizen developer, the no-coder, and the pro-coder. So, what's the difference between these three personas?

All personas are business users. The citizen developer is someone who hasn't got any or much no-code development experience, but it's someone who wants to innovate and is tech-savvy.

A no-coder is someone who has more experience developing without code and knows all about the functionalities of the platform. They can basically build any application with the platform's features.

The pro-coder is a developer who normally codes. They assist the other two personas by answering questions, working on security, and maybe adding some extra functionalities to the platform if their use case calls for some code.

The citizen developer

The citizen developer is a user who wants to innovate or make their job easier by creating an application but has no experience in software development. This could be a business analyst or a project manager, for example. The most important thing is that you are interested in using a new technology to empower your workflow.

So, without any experience in software development, what can these people do? Since they are users with specialist knowledge of the process that they would like to digitize, they are very important for the development of the application. In traditional development, these business users have no role in application development. They basically provide the parameters, and after the programming team has done their work they might do some testing, but that's about it. In no-code development, they become an active part of the application's development cycle.

What would an employee of a company normally do when they have an idea for an application they need for their work? They can usually do two things. The first is using a tool that can help them build an application. In many cases, you see people using Excel to create tools to help them make their work easier or faster. Is this a bad thing? No, but it does create what we call *shadow IT*. Shadow IT is something that usually occurs in larger organizations with a lot of employees, when people start creating, for example, a lot of small systems in Excel or Access. The data in these systems is usually stored on the computer or personal workspace of that user, and this involves some risks: for example, when the person who is maintaining an Excel file leaves the company and nobody else knows about that file, all that data is basically lost. Shadow IT can also easily lead to breaches of data, because everyone can start emailing the Excel file around. These are just some of the examples of what can happen with shadow IT.

Then there is the second option that our employees have. They can go to the IT department and request them to build the software they need. In most cases, it will be put on a backlog and hopefully be picked up as soon as the development team has time for it and the priority is high enough. This often means that employees will be waiting a long time before they get their software and usually are not very involved in the development process. So, they might not get exactly what they need.

When using Betty Blocks, all the apps are created in the company's Betty Blocks environment, which means the IT department knows where it resides and can be a part of the development and maintenance process. This increases the security of the data in the application, and when a developer leaves the company, the application will be easy for the IT department to keep track of.

Betty Blocks allows citizen developers to start creating their own application from the moment they need it. They can start creating wireframes by dragging and dropping a user interface together. Once that is done, they can create the fields that the application will need so that the first prototype will already feel like a first version of the actual application.

The citizen developer will have an active role throughout the development process as well, and they'll learn new skills by taking part in the development cycle. We'll talk about what else they can contribute to the development at a later stage – for now let's move on to the next persona.

To find more information on citizen developers, visit the following URL: `https://www.bettyblocks.com/citizen-developer`.

The no-coder

The second persona is the no-coder. They are usually business users as well, but they can also be dedicated no-coders who work for the IT department. The big difference between the two is that these users are usually very tech-savvy. These people usually learn new computer skills quickly. So, what is the difference between citizen developers and no-coders in the context of the platform? A no-coder can build an application using the platform from start to finish. They have the knowledge to use all parts of the platform to create a fully functional application. Basically, they will do the heavy lifting for the citizen developer when it comes to designing the workflows within the application and help the citizen developer to make their prototype fully functional, taking the citizen developer on this journey while building the application so they fully understand what, and most importantly how, it's built. Since there is no coding involved in the creation of the application, it should be easy for the citizen developer to follow the development and even participate in it.

Does this mean that you don't need any real developers anymore to create applications? Well, we all know that there is a huge need for developers and the business usually needs more from its developers than they can produce in a specific time. This is where no-coder comes in – they can help to create applications faster without the specific need for code-based developers. However, code-based developers should still be involved in the project of creating an application, since they have knowledge the other two personas don't. So, what can they bring to the table?

The pro-coder

Professional coders, or *pro-coders*, still have an important role within the creation of an application, but unlike normal development, they don't have to be involved full time in this cycle. Pro-coders don't need to have full knowledge of the platform itself. They will need to know the basics and how to add functionality to the platform, but it's the no-coders that are the experts on the platform itself.

Pro-coders can add code to the platform, which will allow citizen developers and no-coders to do more with the platform. What does this mean exactly? The whole Betty Blocks platform is made up of a lot of small blocks. Our business users use these to build applications in a no-code way, by selecting or dragging and dropping them. But what if they want to do something with the platform that isn't supported by these standard blocks?

Well, that's where the pro-coder comes in. They can add code to the platform to create or modify existing blocks, which will allow the business users to do more with the platform. For example, imagine that a business user works with a calendar on their web page, but they need this calendar to have an extra functionality for excluding weekends. The pro-coder can modify the existing calendar to support this and then the business user simply drags and drops the no-code calendar block onto their page to gain the new functionality. This is just a simple example, but this can go much deeper.

Also, a pro-coder can help to go over the application before it goes live to check permissions and some specific security requirements that business users might not be aware of.

So essentially, the business users build the whole application, but the pro-coders help them out along the way with some specific use cases. This saves the pro-coder a lot of time, allowing them to focus on other development tasks.

The following diagram shows how the three developer personas can work together within the platform:

Figure 1.1 – Betty Blocks developer personas

So, let's recap what the different personas bring to the development of an application in Betty Blocks.

Citizen developers:

- Innovation
- Wireframing
- Adding basic functionalities
- Designing the workflow
- Testing the application
- Collaborating with no-coders

No-coders:

- Data modeling
- Creating fully functional pages
- Creating workflows

- Connecting to external services (APIs)
- Collaborating with citizen developers
- Testing the application

Pro-coders:

- Adding missing functionality
- Checking security
- Checking permissions
- Testing the application from a developer's perspective
- Assisting with access to external data

Now that we know how the different developer personas can develop using Betty Blocks, let's have a look at what you can actually build with the platform.

Applications that you can build with the Betty Blocks platform

Now we know a little about the different developer personas, but what can you actually build with the Betty Blocks platform? As mentioned before, the platform is cloud based, so it runs on any modern browser. As you might have guessed, you can develop web applications with the Betty Blocks platform. But what kind of web applications, you might ask?

Let's start with the basics of web applications. A web application can be either public or private, which means that you can develop pages that are publicly accessible through the internet with Betty Blocks, but also private pages that require some form of authentication to be able to access them. By default, everything in the platform is set to private, so you don't accidentally expose any private information to the public web.

Also, there is the option in Betty Blocks to build Back Office applications. The Back Office is a part of Betty Blocks and allows you to really quickly set up a basic application with full **Create, Read, Update, Delete (CRUD)** functionality. The interface is fixed, so you can't change the layout, but you do have some options for sorting by columns and fields. Back Office apps are focused on internal use only. The users are different than the ones accessing the frontend – they are basically the same as the builder users, just without the builder permissions. We'll talk about this in more detail later.

All web applications that you create with Betty Blocks are responsive by default, meaning they are accessible from a computer, tablet, or smartphone. This feature comes out of the box with the platform and requires minimal action on the part of the developer to make this happen.

Lastly, what types of applications can you build with Betty Blocks? Let me give some examples:

- An order management portal, where users can access their orders, add new ones, and modify them
- You can build your own custom CRM
- Questionnaire apps, for example, to get information from customers
- You can also build additions to your existing CRM by connecting to your existing CRM and using its data to create a portal for your customers
- You can build an inspection report tool, where people can do inspections on site and add images
- You can build a document management system with a workflow behind it
- You can integrate with almost any API out there to create even more complex applications that use data from internal and external data sources and connect to specific services such as Google Maps, for example

Let's see an example of the ideal areas for deployment of the Betty Blocks platform. If you are planning on building applications within this space, Betty Blocks is the ideal platform to utilize.

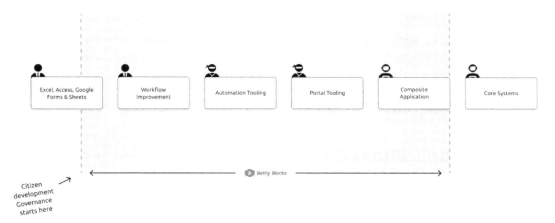

Figure 1.2 – The ideal areas for application development with Betty Blocks

I can give a lot more examples of the kinds of applications you can build with the platform. But basically, the sky's the limit.

So, are there things that are not possible with Betty Blocks? Of course, data visualization over millions of records is not something you would use a platform such as Betty Blocks for, nor creating native applications. Both are possible but are not easily done by citizen developers and require coding knowledge to accomplish. However, here are seven points that do make Betty Blocks stand out:

- **Ease of use (no-code empowerment)**: Users have the tools to develop end-user functionality with their desired look and feel in a simple, fast, maintainable, and non-destructive way.

- **Templates (content) and reusability**: Reuse functional parts across applications to increase the speed of development. No-code is faster than coding, but if you can reuse entire functional parts, you are even faster!

- **Integrations**: Integrate applications with external systems. All connections are secure and based on industry standards and IT retains full control. When you need to connect with the more exotic APIs, there is always the coding escape hatch that lets you build without any restrictions.

- **Builder collaboration**: Applications are always built in (cross-functional) teams. Users can develop and maintain applications together. They have the ability to safely work together on one or multiple applications.

- **Escape hatches for flexibility**: You can build almost anything without a single line of code. But, if needed, you can add code, ensuring you never get stuck. With escape hatches, experienced developers can build any feature in no-code applications by using code.

- **Governance of citizen development by IT**: IT wants business users to build applications themselves, but wants to retain oversight to keep things safe and sound. They want to be able to give citizen developers the right building blocks and environment and govern the entire development and delivery process, and Betty Blocks gives them the platform they need to do this.

- **Scaling citizen development in enterprises**: Enable employees to build their own applications within an enterprise environment while the IT department stays in control of the development process.

But there is so much more that can be done and especially by your business users, which opens up a whole new world. Let's dive deeper into the platform to explore this.

Where do applications on the Betty Blocks platform reside?

As mentioned before, the Betty Blocks platform is a cloud platform, which means it's accessible from anywhere. So, when you create an application on the Betty Blocks platform, where is it hosted? Well, basically there are three different flavors of the platform:

- A public cloud (by default, Azure)

- A private cloud (any cloud operator)

- On-premises

In the public cloud, you can access Betty Blocks anywhere in the world, but can also choose specific areas of the world in which to host your application, including the EU, UK, US, Canada, and many more places. This might be necessary to comply with laws in different parts of the world governing your app and its data. When you start a new application, you can choose where you want it to be hosted. By default, this is always in the Azure cloud.

In a private-cloud setup, Betty Blocks apps are hosted on your own cloud infrastructure, which is maintained and updated by Betty Blocks but owned and operated by you. This is done, for example, if you have a specific cloud preference besides Azure.

The on-premises option means that your instance of Betty Blocks will not be accessible outside your own company. It's also maintained and updated by your own DevOps team (with support from Betty Blocks). You might choose this option if your data is very sensitive.

Summary

In this chapter, we talked about what the Betty Blocks platform is. It's a no-code application development platform that allows a new breed of citizen developer to take an active role in app development. Together with the no-coder and the pro-coder, they can build applications for the enterprise market. Because it's all built in the cloud in an IT-managed environment, it's easy to provide access to any employee in any location.

You can build any kind of web application with the platform since it's based on the latest web standards. Now you understand who can build on this platform and what kind of applications can be built. In the next chapter, we will look at how citizen developers can work with coders on Betty Blocks.

Questions

1. Who can build applications with the Betty Blocks platform?
2. Can you add code to Betty Blocks?
3. What type of applications can you build with Betty Blocks?
4. Where are Betty Blocks applications hosted?

Answers

1. Citizen developers, no-coders, and pro-coders.
2. Yes, pro-coders can add code to the platform to add additional functionality.
3. You can build web applications. These can range from a simple CRUD interface to a whole process-based application.
4. They are hosted in the Azure cloud by default, but the platform is cloud agnostic. Also, you can host applications in almost any area in the world.

2

Collaboration between Citizen Developers and Coders Using Betty Blocks

This chapter is about collaboration between different personas. It focuses on when you build your first larger project, how your team looks, and who you can use in your team for building the application. We'll also talk about how developers can utilize a no-code platform and gain an advantage by using it. Then we'll talk about how developers can help out citizen developers during the development of their applications, and we'll touch on how you can extend the Betty Blocks platform by adding some code to it.

In this chapter, we will cover the following topics:

- Collaboration for building applications
- How experienced software developers can also use a no-code platform
- How pro-coders can help citizen developers
- Adding no-code blocks to the platform with code

By the end of this chapter, you'll have a better understanding of what the ideal team would look like to build a larger application and how developers can support citizen developers while they develop their applications.

Collaboration for building applications

Collaboration is the key to building a successful application. So, who do you need to collaborate with then?

Of course, this also depends on the kind of application you want to build. If you want to build a small application for storing some data that you need from time to time, you can most likely do everything yourself very easily. But let's say that you're going to be very ambitious and want to build a bigger application that you want your whole department or maybe even your customers to use, which supports different types of roles. You also want to have a big process built into the application, which will help the application users to speed up their work. So, what should a successful team look like to build your first application?

First of all, you won't need all of the following people to be able to actually build your application, but they will help you to be more successful, especially if you are doing it for the first time.

So, let's talk about the ideal situation first in which you can achieve the most success when building your first application. We'll assume this is a medium-sized application where you will need some time to build it due to it containing a few workflows. The following are the roles a team could comprise:

- **Project manager**: The project manager keeps track of the general progression of the development and makes sure that everything that developers want to get done is registered so that you can track the progress of the project.

- **Product owner**: The product owner makes sure the value of the project or the product people want to work on is measured. They can also provide the project manager and developers with the user stories they need to start working on their project using the Scrum method.

- **Business users**: The business users can help to define what needs to be developed. Since everything is being built in a no-code way, most business users can fairly easily understand what is being built in Betty Blocks. That way, they can also give continual feedback on what is being developed. Some of these business users might also transform into citizen developers along the way due to interest in the project.

- **Citizen developers**: The citizen developer is usually the person who has the idea for building this application. They will be included in the whole development process and also participate in it. First, they can start with the project manager by working out the general idea and putting that on paper, but they can also start building a prototype in Betty Blocks to showcase their idea. The prototype doesn't need to have any functionality but can be a series of Page Builder pages that gives the other project members an idea of what needs to be built.

  Once the building of the application starts, the citizen developer will work with a no-coder developer to build the application. While the citizen developer is usually becoming familiar with the platform during their first application development, the no-coder does the heavy lifting. The citizen developer is basically participating on all fronts of the development so that they fully understand the process and can transform into a no-code developer.

- **No-coders**: A no-coder is someone who has experience developing with the platform already. They have good knowledge about all the ins and outs of the platform. This user can be someone from the business side or someone working in the IT department. The big difference between the two is usually that the person from the IT department works on different projects with the platform while someone from the business will usually work on a specific project related to their own domain within the company, and they also hold a lot of domain knowledge. The most important factor is that they already have a lot of experience building apps with no code.

- **Pro-coders**: As we discussed, no-coders do most of the heavy lifting during the development of an application, but pro-coders will create most of the logic, help with the finer details of the pages, and help set up API connections with external systems. They involve the citizen developer every step of the way so that their knowledge grows while the application is being developed.

  The pro-coder in this project won't be involved all the time, but they can help out at specific times, for example, when there might be a need to access specific external systems for which something needs to be set up. They can do some testing for security issues, for example, so that there are no leaks of sensitive information.

  The pro-coder is also involved in enabling the citizen developers and no-coders. This means they can create new no-code blocks in the platform using code. These blocks can then be used by the no-coders in a no-code way. So essentially they can extend the functionalities of the platform. This could be a new component in the page builder, for example, or a new action step that allows the user to communicate with an external system or any other backend-related option.

- **IT**: IT will provide the Betty Blocks environment to the users. They will create an application for them with a whole development (DTAP) street so that the users have a safe environment to work in. They can also invite builders to the applications and once applications are completed, they can change the status of an application to production. Production means that the application is live and can be accessed by users other than the developers.

Now that we know which types of personas are useful during the development of an application, let's have a look at how pro-coders can utilize a no-code platform and what advantages they can take from this compared to normal coding.

How can developers also use a no-code platform

Is a no-code platform also useful to a pro-coder (developer)? The short answer is yes, of course. So, what advantages can a no-code platform such as Betty Blocks have for a pro-coder?

The most important advantage of a no-code platform is the speed for developers. Let's say they have to develop a portal for their company. Usually, a portal is something pretty straightforward. It usually allows the company's customers to log in to their portal and access all kinds of information. For example, they might be able to see the orders that they have placed with the company and also the invoices that accompany them. They can view these and even download the invoices for their own use, and they can also see the status of their orders. All this information can come straight from the company's ERP system. Betty Blocks integrations allow you to quickly connect to these systems and receive and send data to them. This way, Betty Blocks can utilize this information to show it to the users of the portal.

Building this portal is something that can be done very quickly with a no-code platform because, as discussed, it not only integrates easily with any external system through APIs but also allows the developers to set up a data model (database) quickly and easily within minutes that will hold the information specific to the customer who is logging in to the portal. The Page Builder's drag-and-drop functionality allows them to easily set up the portal. Security is already taken care of with the login flow built into Betty Blocks, and even if they need to use **single sign-on** (**SSO**) because their customers are already using that for other applications, this can be integrated quite easily by grabbing one of the templates that already has SSO pre-configured.

Since the developers don't need to set up a specific development environment and configure the libraries that they might need for the project, there's no need to set up a database or do any coding. They win back up to 50-75% of the time it would take to build a portal. And as mentioned before, if there is anything that might be very specific and not supported out of the box by the platform, they can add some code to still be able to finish the portal to their liking, but we'll talk more about that in the last two sections of this chapter.

Since the backlog of development teams is pretty big in most companies, utilizing a platform such as Betty Blocks could be very advantageous to reduce this backlog. It might not suit every use case for developers, but it can speed up their development by up to 75% and also make certain development tasks a lot more fun because it's much easier and faster (and therefore enables faster results) to do with essentially the same control as you have with code.

Pro-coders can gain a lot by using a no-code platform when the use case fits. But one of the most important jobs a pro-coder has is helping out the citizen developer and no-coder. Let's explore this in the next section.

Pro-coders can help citizen developers

Citizen developers are new to building their own applications and need some help from time to time. We've already mentioned in the *Collaboration for building applications* section the other roles that can help them out, but let's zoom in more on what coders can do for them specifically and why it will also save the coders time in the long run by helping out citizen developers.

Of course, it will not only be the citizen developers who will need help sometimes, but also the no-coders. Basically, what goes in this section for the citizen developer also goes for the no-coder since those two will work together on their project usually, or either one of them will work alone on a project, but on some projects, they might need some help from a pro-coder, which is what this section is about.

So, let's talk about where experienced developers (pro-coders) can help out the citizen developers. Betty Blocks has many templates for making more complex things a lot easier, but even then, some of the details might still be a bit intimidating for citizen developers the first time they encounter them. For example, they start a new application that needs to integrate with the company's SSO. For the most common SSO, there is a template available, but you still need to enter some details that allow you to connect to the company's SSO; otherwise, everyone can use this. A pro-coder can help the citizen developer to set this up. It will only take a few minutes, but it will make it a lot easier for the citizen developer. It also allows the pro-coder to do a few security checks at this point.

Another use case might be that the developer wants to retrieve data from an existing application or database. Pro-coders usually need to set up the credentials for the citizen developer because the citizen developer won't have access to the backend of the existing application or database. Betty Blocks already has a whole library of APIs for this that can be downloaded, integrated, and used straight out of the box. Even here, the pro-coder can help to enter the login details of the API to set up the API together with the citizen developer. And maybe in some cases, if the system they want to connect to is not a very common one, but the pro-coder has the documentation on how to set up an API connection, they can create a web service in Betty Blocks for the citizen developer. Once this is set up and functional, they can push it to the block store so that the citizen developer API can be used by anyone in the organization in their next application without any effort.

A pro-coder can also help when the citizen developer is building a page in the page builder and wants to do something that might not be supported out of the box. The page builder allows the pro-coder to create a ReactJS-based drag-and-drop component in code by adding external libraries to new components for even more functionality, which can then be used by the citizen developer. Betty Blocks has a **command-line interface** (**CLI**) that can help with supporting this. We'll talk more about this in the next section. But it enables the pro-coder to add almost any code functionality to the platform in a no-code way.

The same goes for logic in Betty Blocks, which is called actions. Actions are usually backend functionality, but in Betty Blocks, you can design these in a no-code way by dragging and dropping them. These actions are like a visual workflow and have steps and conditional logic. Each of these steps is written in Node.js, which is a backend JavaScript language used in Betty Blocks. Here the pro-coder also has the option to add code to the actions by using the CLI, just like with the page builder. So almost anything the pro-coder can think of, they can do by adding code and enabling the no-coder to do more.

Adding no-code blocks to the platform with code

Betty Blocks is a no-code platform, but does this mean you can't add any code to it? No, you can add code to Betty Blocks. So, what can you do with code in Betty Blocks?

There are three places where you can add code to Betty Blocks:

- **Actions**: Here, you can use code to extend the logical part of Betty Blocks. For example, there could be a situation where you want the logic of Betty Blocks to do something that it normally can't do, or you want it to work slightly differently. You can create new code or edit existing code to make changes to the actions. Then, a no-coder can use them in a no-code way on the platform.

- **Page builder**: Here, you can add new components or modify existing ones. For example, if you want to add a new component to display a calendar, you can create it in code and add it to the platform so that a no-coder can use it in a drag-and-drop way.

- **Endpoints**: Here, you can add code directly to the platform. You can write HTML, CSS, and JavaScript to create pages, just like you would normally do when creating a page in a code editor.

For actions and the Page Builder, pro-coders and experienced developers can use the Betty Blocks CLI to add their code to the platform. The CLI allows pro-coders to use their favorite code editor and create any code to add to the platform. For actions, this is done with Node.js, and for the page builder, this is done with ReactJS. Betty Blocks also has what we call prefabs, which allows pro-coders to create an interface on their code so that the code becomes a no-code component. This allows citizen developers and no-coders to be able to utilize these new components without having to understand the underlying code. And once this piece of code has been written and uploaded to Betty Blocks, the pro-coders can share this new no-code block with their whole organization so that they don't have to rebuild it for each new application.

If you want to have a look at the Betty Blocks CLI, you can visit `https://github.com/bettyblocks/cli`. Here you can read about the CLI and install it. We'll be talking about the CLI in the chapter *Pro-Coder Features* of this book.

Just remember this part of the platform is only for experienced developers and allows them to extend the platform without being constrained by it if they have specific wishes. But it's not a mandatory part of the platform that you need to understand if you want to be able to build applications. You can build so many applications without having to add custom code to the platform, but it is important that you at least know that the possibilities are there.

Summary

In this chapter, we learned about the collaboration between different personas to develop an application. We assumed that it was going to be a somewhat larger application, so you needed a small team to make this successful. The personas in this team were a project manager, business users, citizen developers, no-coders, pro-coders, and IT.

The added value of a no-code platform such as Betty Blocks for a pro-coder is another topic we talked about. Pro-coders can win a lot of time back by utilizing the platform for specific use cases, such as portals and internal CRUD applications. This could increase their development speed by up to 75%.

We also learned that pro-coders can help citizen developers and no-coders. They can add extra no-code functionality to the platform by using code so that other developers can do more with the platform. They can also help by performing security checks during development.

Lastly, we spoke about adding code to the platform to create new no-code functionalities. This should help developers to do more with the platform or to perform specific tasks that otherwise might not have been possible.

In the next chapter, we'll have a look at governance. Betty Blocks has its own governance section called My Betty Blocks. This allows you to govern all your applications and users.

Questions

1. What would an ideal team look like to develop a new application?
2. What kind of code can you add to Betty Blocks?

Answers

1. A project manager, business users, citizen developers, no-coders, pro-coders, and IT.
2. You can add new components to the page builder and add code for new actions, allowing users to create more logic.

Part 2:
First Steps on the
Betty Blocks Platform

This part is all about taking your first steps on the platform. You will learn how to create an app, how to invite users to your app, and how to use a template to explore the platform. This part has the following chapters:

3

Governance on Betty Blocks

Now, we've arrived at the point at which we will cover the Betty Blocks platform in depth. We'll start this chapter by talking about governance in general and how this applies to the Betty Blocks platform. Then, we'll get you started by creating your account on the platform, which will enable you to follow along with the book – you can try out everything that we will describe.

In the next section, we'll talk about My Betty Blocks. This is where all the governance of the platform takes place. We'll dive into organizations, what they are, and what they do to help you in the development of your applications.

We'll create our first application and show you how to do that. Once you've created your application, we'll take you through all the options that are available for applications and how you can invite new admins or users to your applications. I'll also show you how you can create a sandbox of your application and explain what exactly sandboxing is and means and how useful it is to use sandboxing.

Lastly, we'll talk about templates and how you can utilize them so you can get started with your application more quickly.

We will cover the following topics:

- What is governance on Betty Blocks?
- Organizations – the primary layer for your apps
- Applications within your organization
- User management
- What is sandboxing?
- Using application templates

By the end of this chapter, you'll have learned how to govern your applications in Betty Blocks, manage your users within your organizations and applications, and what sandboxing is and how to use it.

What is governance on Betty Blocks?

The whole process of creating new applications starts with governance. The reason for this is that for every application that is created, you want to know about it. Otherwise, there will be a wild growth in applications and no monitoring of what those applications do or what they are being used for. This will lead to shadow IT and the risk of losing data since nobody knows where this data has been stored.

This is why the governance of Betty Blocks usually lies in the hands of the IT department or the innovation department. This way, they can facilitate and manage the creation of new applications and **Development, Testing, Acceptance, and Production (DTAP)**-streets, merge work that has been done in production, and invite new developers to applications. This way, the citizen developer and the non-coder don't have to worry about these things, especially within larger companies, as everything is managed from one central point.

So, where does governance in Betty Blocks take place? It all takes place at My Betty Blocks (`https://my.bettyblocks.com`). Here, you can do everything related to the things mentioned in the previous paragraph. We'll go through all of this in detail so that you understand what all of this does and how it can help you get the best out of the platform.

Setting up your account

You might be completely new to Betty Blocks or no-code solutions, so you'll have no Betty Blocks account yet. Let's get that fixed first so that you can follow along with this book more easily. First, you'll need to open your browser and go to `https://www.bettyblocks.com`.

This will land you on the home page of Betty Blocks as shown. In the top-right corner and center of the page, you'll see buttons called **Start free trial**. This trial is accessible for at least 30 days. It's not automatically canceled after 30 days, but the marketing or sales department at Betty Blocks controls the overall length of your trial.

Figure 3.1 – Betty Blocks home page

Click on that button. This will take you to the registration page for the platform. Fill your information in here to get started:

Register

Get started by creating an account first

FIRST NAME *   LAST NAME *

BUSINESS EMAIL *

A valid email is required to verify your account

PHONE NUMBER *   COMPANY SIZE *

Select company size   ⇕

LEVEL OF EXPERIENCE *

Do you have development experience?   ⇕

To make sure you get the best and most complete Betty Blocks experience, we need to send you additional content and materials. We will not use your address for any other purposes.

I agree to receive Betty Blocks experience updates and resources

I agree to Betty Blocks' Terms and Conditions and Privacy Policy *

Register

* By submitting this form, you give consent to Betty Blocks to store your information above to generate your account.

Already have an account? Sign in

Figure 3.2 – Registration page

Afterward, you should receive an email to confirm your registration. Once you have, you'll be asked to choose a password and you should be logged in to my.bettyblocks.com.

You're now registered on the platform and ready to get started building your first application.

Introduction to my.bettyblocks.com

Now that you are logged in to My Betty Blocks, let's quickly introduce where you have landed in My Betty Blocks.

Once you log in to My Betty Blocks, you land on the dashboard, from which you can quickly access some of the pre-built templates offered by Betty Blocks. It also gives you quick access to the learning section of Betty Blocks, which offers different types of videos for learning on the platform, and also offers a tour to provide a quick overview of the platform.

Next, we'll have a look at what organizations are and how you can manage them as an admin.

Organizations – the primary layer for your apps

Organizations are the basis of your applications. Every application in My Betty Blocks resides in an organization. An organization can hold multiple applications, but it also decides where your application will live in the world. Betty Blocks is hosted on different servers in the world. Since laws differ across countries, you may want to host specific applications in specific countries to make sure your application and the data in that application apply to a particular country's set of laws.

Once an organization has been created in a specific zone – a zone is a specific area in which your application lives (i.e., the EU, the US, or Canada) – it will always create its applications there as well. As talked about before, Betty Blocks runs by default on Azure in the public cloud, but it can also run on other cloud providers. In the public cloud, you have the choice to deploy your applications to different zones – for example, in the EU, the UK, the US, or Canada. Depending on which zone your organization is created in, your applications will be automatically deployed there.

When you created your trial account on the platform earlier, the platform also created an organization for you. Log in to the platform and let's have a look at this organization. After you log in, you should end up on the dashboard in My Betty Blocks. On the left, you should see the vertical menu. Click on the icon that looks like a building: this should take you to the organization overview page. The following screenshot shows an example of what this looks like:

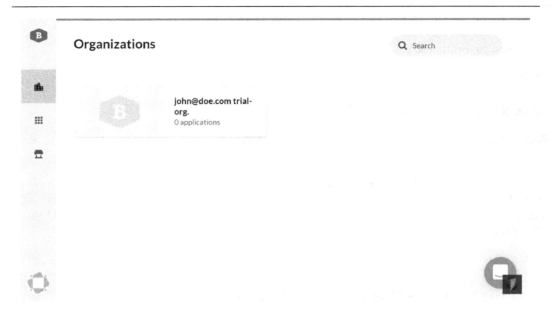

Figure 3.3 – Organization overview

Now, let's click on the organization panel that you can see. This should open up the organization itself. You should see a screen like this:

Figure 3.4 – Inside the organization

Organizations have the following options that you can interact with:

- **Settings**
- **Members**
- **Create application**

Settings for organizations is not very extensive, but you can change the name of your organization here and view the zone your organization is part of.

The **Members** option shows you all the members of your organization. You can either invite users directly to your organization, or you can invite users to an application that is part of this organization. In both cases, the user is added to the organization members as well. There is a distinction between types of users in organizations: you can either just be a member or be an admin. Being a member means that you can only see the applications that you are added to. If you don't have access to any applications, you will just see an empty organization. The user you are logged in with now should be an admin. Being an admin allows you to create applications in the organization and also invite users to these applications. It allows you to merge application sandboxes. We'll dive deeper into this in the *What is sandboxing?* section later in the chapter.

Members are always invited to join an organization by using their email addresses (see *Figure 3.5*). Betty Blocks also has options for larger organizations to add members automatically through their identity provider. This allows organizations to quickly add and remove members using their identity provider without having to manage them in My Betty Blocks:

Figure 3.5 – Organization members

With the **Create application** button, you can create new applications in your organization. Once you click on this button, it will open the application template overview. Here, you can select a template or click on the **Start from scratch** button. This will create an empty application without anything preconfigured.

Let's create an application so that we can have a look at the options there are for applications as well. Click on the **Start from scratch** button. A dialog box will open, in which you can enter the name of your application. Enter any name that you like:

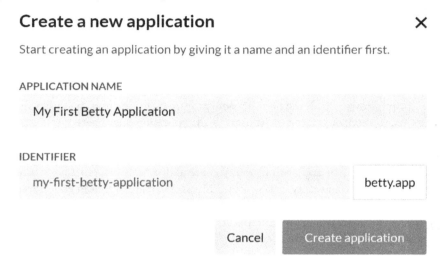

Figure 3.6 – The new application dialog

As you can see, the identifier is also filled in with your application name. The identifier needs to be unique, so if it's not unique, you need to change it so that it is unique within the platform. Once you are finished, click on the **Create application** button.

Applications within your organization

Creating your new application might take a few seconds or minutes, so wait for that time. Once it is done, let's have a look at our new application in My Betty Blocks. You should have been redirected to the organization overview in which you can see all of your applications. It looks as follows:

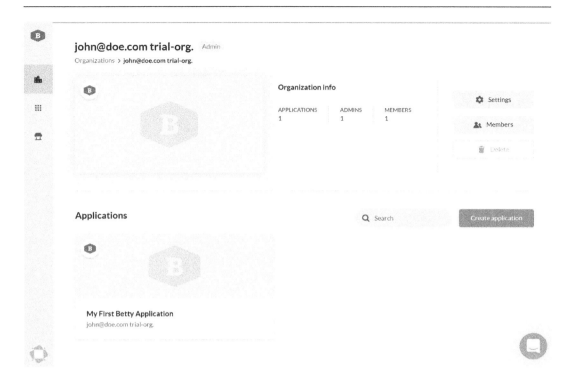

Figure 3.7 – Organization overview with one application in it

When you hover your mouse over the application, you should be able to click on it, and it should take you to the application overview. Be careful not to click on the **Open** button because this will take you to the development environment for the application (we'll go there in the next chapter).

Now, we are in the overview of your application, in which you can manage your application settings and invite users to your application. Let's start with the settings of an application. You can access the settings of an application by clicking on the three dots on the right-hand side:

Figure 3.8 – Application overview – settings menu

Here, click on the **Settings menu** option. This should take you to the application settings for your application. Here, you have six different settings tabs:

- **Options**
- **Web options**
- **Hostnames**
- **Mail**
- **Invitation**
- **Template**

Let's discuss each setting in detail.

Options

Here, you can set up most of the basics of your application for My Betty Blocks. You can add an image for this application so that it shows up in your application overview instead of a gray Betty Blocks logo. Next, we have the name of the application. You will have already named your application during the creation of your application, but you have the option to change it here.

You also have the option to change your identifier since this needs to be unique and also changes the URL of your application. Changing this might have a considerable impact on your application, so please choose your identifier carefully:

Options

Manage name, image and settings to customize your application.

ICON AND BACKGROUND

Maximum upload size: 5 MB

NAME

Academy

IDENTIFIER

online-academy   .bettyblocks.com   ✎ Edit

Your application will be disabled when changing the identifier, this can take up to 20 minutes.

ORGANIZATION

Online Academy   ⌄

Advanced

Logging of page and action debugging and other advanced options.

LOGGING

MANDRIL API KEY

2FaLS2HeZIv5Fues4CR-4g

USABILLA SCRIPT IDENTIFIER

Copy paste the id. For example: https://w.usabilla.com/12345678abc.js, the id is 12345678abc.

Figure 3.9 – Options

If you are an admin for multiple organizations, you can also switch organizations within your application, and move from one organization to another.

Next up are the more advanced options, starting with logging. As we start building our application, we might want to log some of the actions taking place in our application. Logging can be quite useful for debugging, but it also slows down your application. So, with this option, you can turn off all the logging options across your whole application with just one click. This is especially useful when you are deploying your application to production later.

The private data mode is a very important option. This is turned on by default when you start your application. What does this mean? Your application will have a database with data inside it. By default, this data won't be accessible if you are not logged in to the platform itself. Basically, it protects your data by not setting the data as public. This allows your application builders to build applications without the risk of exposing data to the outside world before the application has been officially released and checked by, for example, IT or an experienced developer. When this option is not enabled and you're visiting the frontend of your application, the system will recognize that you're not logged in to the platform and redirect you to the login page first.

Web options

Web options has been specifically designed for the part of Betty Blocks that allows you to build web pages:

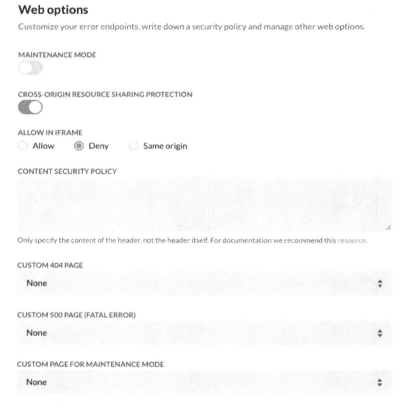

Figure 3.10 – Web options

The **MAINTENANCE MODE** option can turn on maintenance mode for your applications. This will display a maintenance message for all users of the application so that nobody can use the application during maintenance. This goes for both the development and the runtime (the frontend) part of the application – be aware that this is only for the older IDE in the platform.

CROSS-ORIGIN RESOURCE SHARING PROTECTION is a very technical option. Essentially, this option allows you to share resources across multiple websites – so if, for example, you want to use things you have built within Betty Blocks on another website or application, this option needs to be enabled.

ALLOW IN IFRAME has multiple options: allow your application to be embedded in iframes from external applications, deny your application to be embedded in iframes from external applications, or allow your application to only be embedded in iframes from the same application. This can be useful if you are building an application that needs to be part of another application and be hosted in an iframe within that application.

CONTENT SECURITY POLICY lets you specify the content of your content security policy headers. This will add specific headers to your pages for security. This only works in the custom endpoints of Betty Blocks, not in the page builder, so if you are building with the page builder, you don't have to worry about this option.

The three options for setting your **404**, **500**, and **MAINTENANCE MODE** pages are also specifically for the endpoint pages of Betty Blocks. You set these here based on pages that you've created in the application so that those pages will show up instead of the standard 404, 500, or maintenance pages of Betty Blocks. The page builder has a new setup for this, which is handled in the Betty Blocks application.

Hostnames

Hostnames allows you to use your own domain for your application. By default, your application will have a name such as `<identifier>.betty.app` for the application frontend. To change this to your own domain or subdomain, you can use a hostname to set this up:

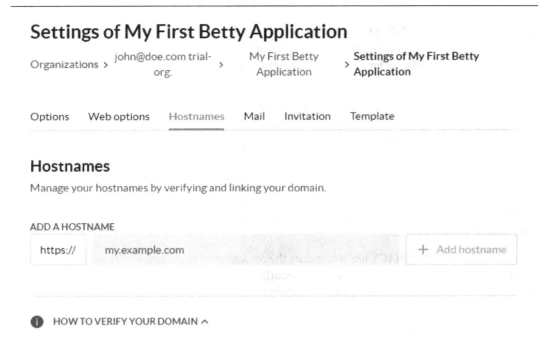

Settings of My First Betty Application

Organizations > john@doe.com trial-org. > My First Betty Application > **Settings of My First Betty Application**

Options   Web options   Hostnames   Mail   Invitation   Template

Hostnames

Manage your hostnames by verifying and linking your domain.

ADD A HOSTNAME

| https:// | my.example.com | + Add hostname |

ℹ HOW TO VERIFY YOUR DOMAIN ∧

Figure 3.11 – Hostnames

Here, you can set up your own domain and then upload a certificate for your domain. All domains for Betty Blocks are secure (SSL) by default, so you always need to upload and send an SSL certificate to Betty Blocks in order to get your domain working. This part of the process is handled by Betty Blocks support – installing the certificate, for example. You can read the instructions about that on this page as well.

Mail

The **Mail** options for Betty Blocks are configured here. There are three variants that you can choose from:

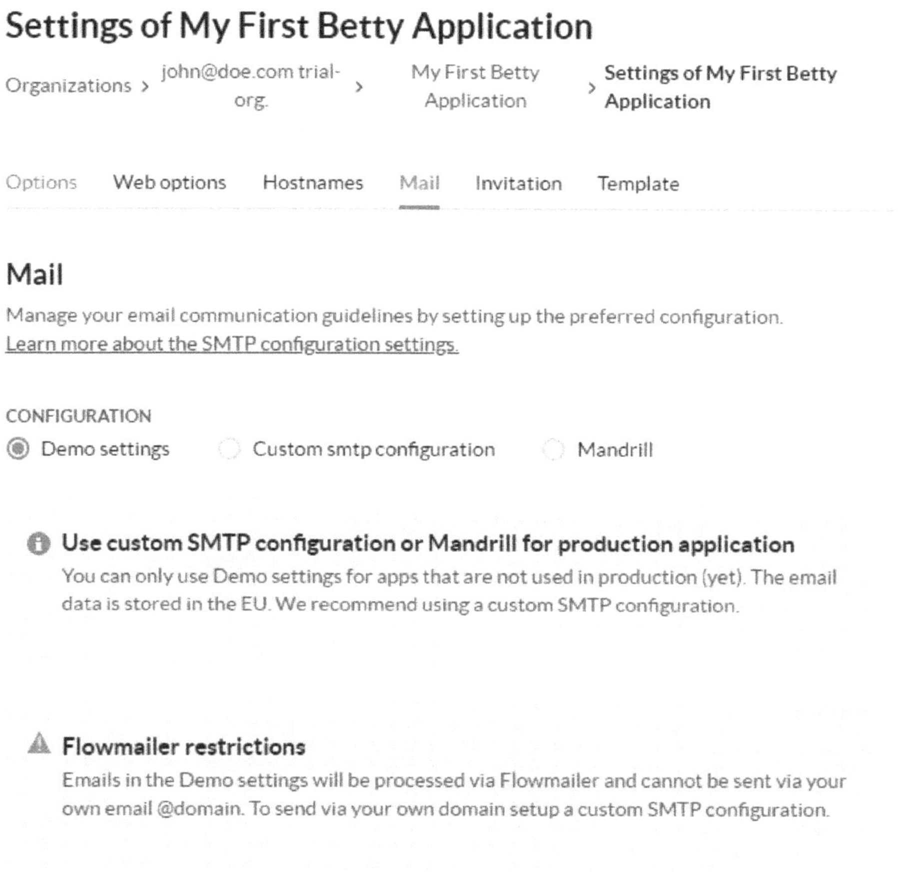

Settings of My First Betty Application

Organizations > | john@doe.com trial-org. | > | My First Betty Application | > | Settings of My First Betty Application

Options   Web options   Hostnames   Mail   Invitation   Template

Mail

Manage your email communication guidelines by setting up the preferred configuration.
Learn more about the SMTP configuration settings.

CONFIGURATION

◉ Demo settings   ○ Custom smtp configuration   ○ Mandrill

ⓘ **Use custom SMTP configuration or Mandrill for production application**
You can only use Demo settings for apps that are not used in production (yet). The email data is stored in the EU. We recommend using a custom SMTP configuration.

⚠ **Flowmailer restrictions**
Emails in the Demo settings will be processed via Flowmailer and cannot be sent via your own email @domain. To send via your own domain setup a custom SMTP configuration.

Figure 3.12 – Mail options

The first **Demo settings** option is the standard one that is activated when you have just created a new application. This option allows you to send emails from the `betty.app` domain (for example). This is the only extension you can use by default. This one allows you to send emails by using `noreply@betty.app` as an email address. You can change the `noreply` part of this email address, but `@betty.app` is the only extension that you can use. This prevents users from accidentally emailing from specific accounts during the development of their application.

For production applications, we have **Custom smtp configuration** and **Mandrill**; these two options allow you to use your own domain.

SMTP needs to be set up first. You'll need to configure the SMTP settings for your domain so that you can send emails from your domain first. Once you've filled in these settings, you are good to go.

For **Mandrill**, you need to have a Mandrill account or use the default one that Betty Blocks provides. Mandrill is an external transactional email service, so, if you are going to use a lot of bulk emailing, this might be a good option. You can also use Mandrill to verify your own domain. This will allow you to send emails from your own domain as well, just like with SMTP.

If you are not going to be bulk emailing – just sending emails to external parties and internal employees – then SMTP is the best choice since you use your own email server to send emails.

Invitation

Invitation handles the email that is sent to users when they are invited to the application. This is purely for the backend part of the application, where you develop your application and have access to the back office of the application. These users will become part of the Betty Blocks ecosystem. The frontend users who are invited to the application are only a part of that specific application and have nothing to do with Betty Blocks. We'll dive deeper into this in the chapter where we talk about the page builder:

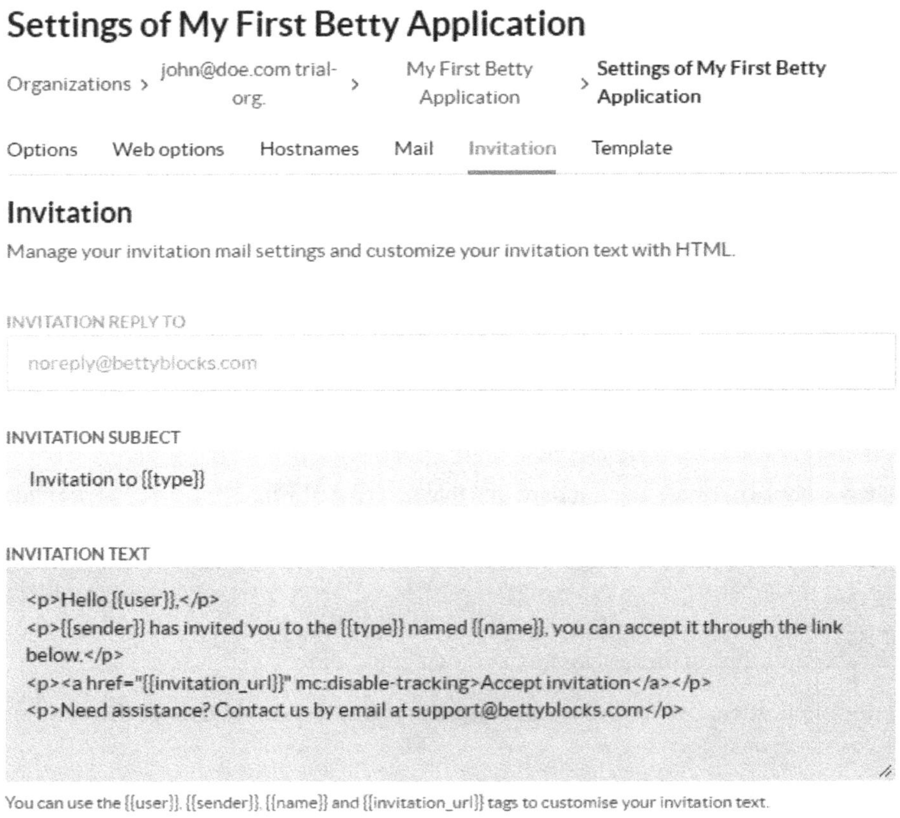

Settings of My First Betty Application

Organizations > | john@doe.com trial-org. > | My First Betty Application > | **Settings of My First Betty Application**

Options   Web options   Hostnames   Mail   Invitation   Template

Invitation

Manage your invitation mail settings and customize your invitation text with HTML.

INVITATION REPLY TO

noreply@bettyblocks.com

INVITATION SUBJECT

Invitation to {{type}}

INVITATION TEXT

```
<p>Hello {{user}},</p>
<p>{{sender}} has invited you to the {{type}} named {{name}}, you can accept it through the link below.</p>
<p><a href="{{invitation_url}}" mc:disable-tracking>Accept invitation</a></p>
<p>Need assistance? Contact us by email at support@bettyblocks.com</p>
```

You can use the {{user}}, {{sender}}, {{name}} and {{invitation_url}} tags to customise your invitation text.

Figure 3.13 – Invitation

Here, you have the option to make changes to the email that is being sent out to the user who is being invited. By default, it's the standard email that Betty Blocks sends, but if you would like to personalize this email, you can do that here. There are also a few options that you can use, which are shown between curly brackets. For example, {{user}} will replace this part of the email with the name of the user. Let's see what all of these do so that you know how to use them:

- {{user}}: Shows the user's name
- {{type}}: Shows the type of the application
- {{sender}}: Shows the name of the user who sent the email
- {{name}}: Shows the name of the application
- {{invitation_url}}: Shows the URL of the invite so that the user can accept the invitation

With this, you should be able to modify your email to suit your own.

Template

Template allows you to convert your application into a template. Once you are done with your application and you want your application to function as a template for other users, you can use this to set it up as a template. When you do this, only people within your organization can use this template. Not all users within the platform will be able to see it.

You can fill in all the fields here and you're ready to go:

Settings of My First Betty Application

Organizations > john@doe.com trial-org. > My First Betty Application > **Settings of My First Betty Application**

Options   Web options   Hostnames   Mail   Invitation   Template

Template

Make your application's design reusable by converting it to a template.

TEMPLATE NAME

Fill in the name of the template

TEMPLATE DESCRIPTION

Write a description about your template

PREVIEW URL

Fill in a URL to preview the application

DOCUMENTATION URL

Fill in a URL of the documentation

Figure 3.14 – Template

These are all the options you can set for your application. You don't have to worry about most of them when you first get started with building your application, but it's good to know that these options are there for you. Let's talk more about user management now.

User management

As mentioned before, there are two different types of users: application users and organization users. Eventually, all application users will end up as organization users as well because an application is always part of an organization. We've already talked about organization users, so let's dive deeper into application users now. In the next screenshot, you can see an example of a user who has been added to an application:

Figure 3.15 – User overview

An application user comes in two different types: an admin or a member. The admin is essentially a builder of the application. This user has full access to all the elements of Betty Blocks that allow them to build and make changes to an application, such as the page builder, the models, the actions, and the web services. This user also has access to the data API, which means that you can access the data in your application from the frontend of your application while it's still in development. Frontend users don't have this option (don't confuse frontend users and members here).

Members are users that have access to a small part of the application's backend. The only thing they have access to is the back office. This is a special part of the application that allows users to quickly interact with and manage the data inside the application. The back office is something that can be set up very easily and quickly, but still has all the power of the CRUD user roles that all the other users also have, which means you can also use this as your CMS or admin environment. What exactly does this mean then? Why would you use this user at all?

This has to do with where the Betty Blocks platform comes from. When Betty Blocks was first developed, it only had what we now call the back office. It was a simple CRUD interface that allowed users to quickly set up an interface to interact with data. There was no frontend in Betty Blocks at that point (we're talking about 2013-2014 now). The back office still remains a powerful tool because of its ease of use, although it is something that is a less prominent part of the platform nowadays. We'll still cover it in this book since it's so easy and powerful. You can create internal applications with this back office for your employees.

So, what happens if you delete a user? If you delete a user from My Betty Blocks, the user doesn't have any access to this application anymore, but the actual user data is not deleted from the application because that would mean that you would lose all the reference data for that user as well. The user can't see the application anymore in My Betty Blocks and can't log in anymore, but the data remains intact. Of course, you can write an action in your application that removes all user data as well if you wish to, but it's not done by default.

Users also need to be invited to each sandbox of an application separately. This means that if you are invited to one of the sandboxes, you don't automatically have access to all the other sandboxes. Let's talk about sandboxing in more detail now.

What is sandboxing?

Betty Blocks has an option called sandboxing for your application. If you have never programmed before, this won't immediately make sense – so, what is it exactly?

Sandboxing is a way of splitting up your application into multiple applications, so to speak, which allows you to develop your application without changing the production application. Now, you might be thinking, but that's exactly what I want to do – change my production application. Of course, that's the goal, but if you are making a change to your production application and someone visits your application at that point, they might get a broken page, for example – and you don't want that. You want users to be able to keep doing what they are doing while you work on your application because it might take several hours or maybe even a few days before you are done with making your changes and testing them. This is where sandboxing comes into play.

When you sandbox your application in My Betty Blocks, you basically make a copy of your production application. This copy is still connected to your production application and each sandbox has its own database, so you never mess up your production data in those sandboxes. In this copy, you can make changes, which you can also test in this same copy that you've made. Once you are happy with your changes, Betty Blocks offers you an option called **merging**. This means that you can copy over all of the changes to the application that is connected to your copy. Let's see how this works.

Here you can see an example of what your current application should look like:

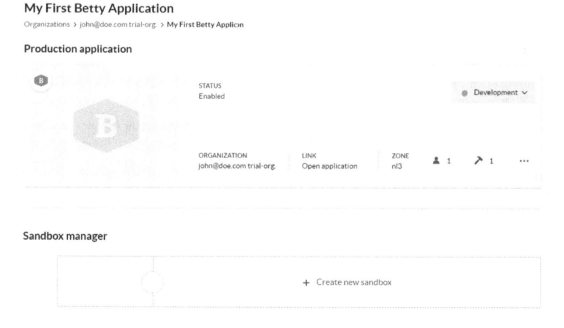

Figure 3.16 – A production application without a sandbox

Underneath your application, you will see a box with the **Create new sandbox** button in it. Click on it, which should open up a dialog that allows you to name your sandbox. Because it will be a completely new application, you need to give it a new name. Let's name it `My First Betty Acceptance`. The identifier of your sandbox will still have part of the original identifier, as you can see in the dialog, and it will copy the name of your application there as well. Since this will make the identifier extremely long, we'll change it to just `acceptance`. Let's click on the **Create new sandbox** button. This will take a minute to run.

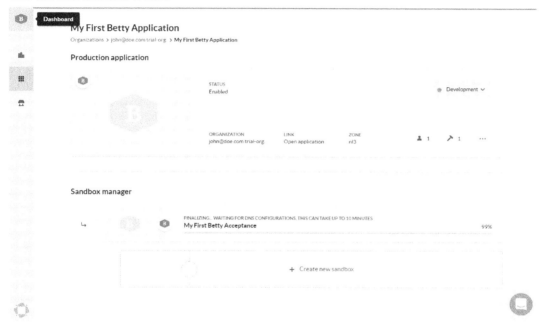

Figure 3.17 – Creating a new sandbox

So, why the name acceptance? In most cases, you create four sandboxes for your applications, which are called a DTAP-street in the programming world. As you can see, we already have our production environment, so now we are creating an acceptance environment. The acceptance environment is usually used to test the latest changes by a specific group of users, for example. The test environment is used by developers or specific testers to test the application, and then there is the development environment, in which the actual development is being done. This way, testers can test without having to worry about sudden changes from the developers, and changes can be pushed slowly from sandbox to sandbox until they are production ready.

Now that your acceptance sandbox is ready, it should look something like the following screenshot:

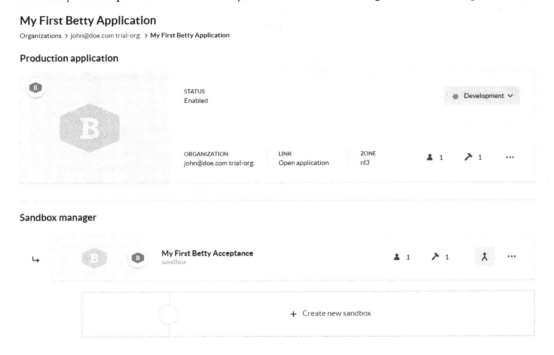

Figure 3.18 – Application with one sandbox

Let's create the whole DTAP-street now – name the sandboxes:

- Name: `My First Betty Test`
- Identifier: `Test`
- Name: `My First Betty Development`
- Identifier: `Development`

Every time you create a new sandbox, it will take a minute before you can create a new one again, so this might take a short while. Once this is done, it should look like this:

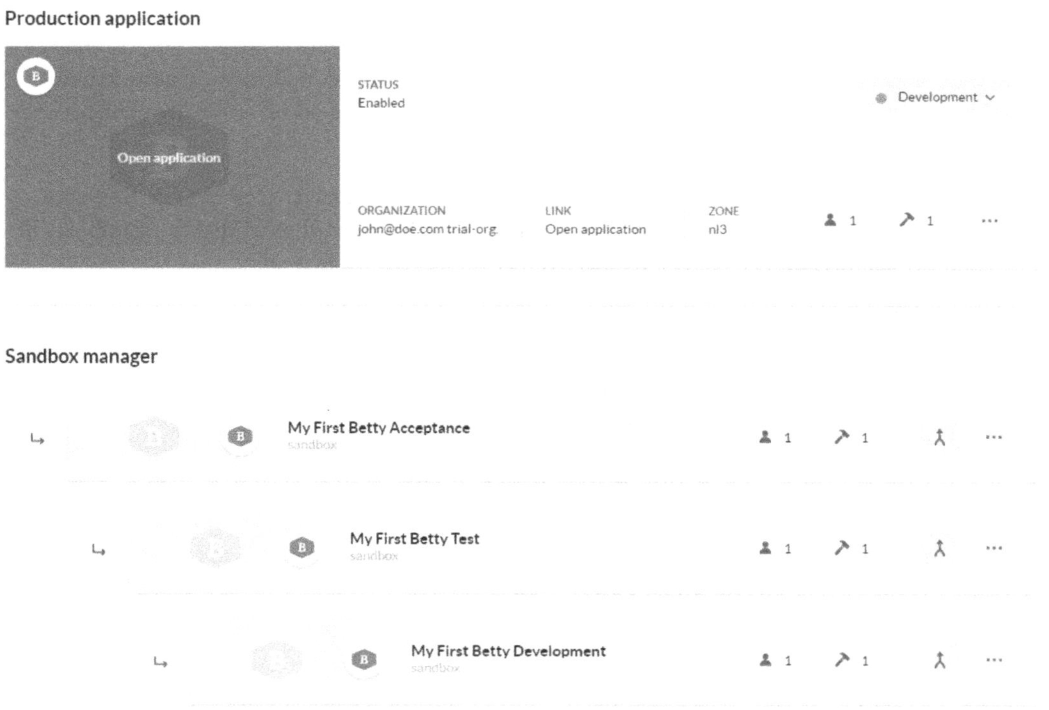

Production application

STATUS
Enabled

Development ∨

ORGANIZATION
john@doe.com trial-org.

LINK
Open application

ZONE
nl3

👤 1   ↗ 1   ⋯

Sandbox manager

My First Betty Acceptance
sandbox

👤 1   ↗ 1   人   ⋯

My First Betty Test
sandbox

👤 1   ↗ 1   人   ⋯

My First Betty Development
sandbox

👤 1   ↗ 1   人   ⋯

Figure 3.19 – Application with a full DTAP-street

Now you've set up a full DTAP-street for your application – so if you made a change to your application, how would you push this to the production application?

You can do this by clicking on the merge button (人). As you can see, each application has a merge button and merging is always done from bottom to top. One of the things that sandboxing does is disable development in all applications except for the lowest one, so you can only develop one application, which is always the last one you created. If you were allowed to make changes to the other applications, the merging would become a mess, and your applications wouldn't match anymore. Of course, you can manipulate data in any of the applications. Each application has its own data, so if you make a change to your data in one application, it won't affect the others.

Now, let's say you've added a page to your application and you then click on the merge button for your development application. You'll see an overview of all your changes after clicking on the button, which you can start to merge to the next application (in this case, the test application). Once they have been tested in the test application, you can merge them from the test environment to the acceptance environment, and then from acceptance to production. This sounds like a lot of work, but it does reduce the chance of breaking your production application because everything goes through a test cycle as well.

The last thing about merging: the only users that can merge are organization admins. Developers of applications can't merge – if you are both a developer and an organization admin, you can merge, but you'll need to have both admin roles. Otherwise, you'll have to ask your organization admin to merge the changes for you. This way, you ensure that you have one person making the changes and another person applying them and making sure things get tested.

In a later chapter, we'll do an actual merge since we haven't done any actual development in the platform yet – we'll get to that later. We'll merge our changes to an underlying sandbox so that the sandbox will also have all the changes that have been made to the development environment.

Now, the last thing we'll talk about in this chapter is how to use templates. Let's check that out in the next section.

Using application templates

Betty Blocks has a library of standard templates available. All of these templates are free to use for anyone who has access to the platform. As mentioned before, you can also create your own templates once you've finished building your application or template. You can publish it to the template section of My Betty Blocks, where anyone who is part of your organization can use it. Let's have a look at the standard templates first:

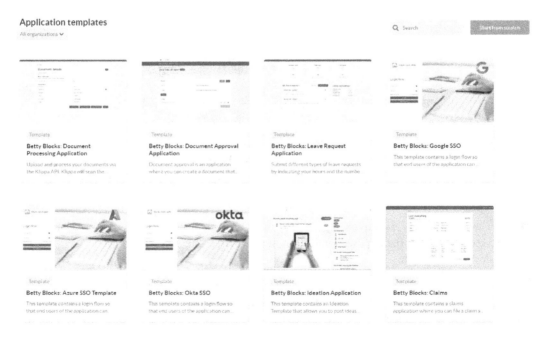

Figure 3.20 – Standard templates in My Betty Blocks

As you can see, these are a few standard templates that Betty Blocks offers. There are templates here that can help you kickstart your application, and there are templates that are already complete applications. The ones that are there to kickstart your applications include the Google SSO template, for example, which has Google SSO built in, so you don't have to build it yourself anymore. There is no actual application with a workflow besides the login in this template – that is the part you'll have to build yourself. However, it does save you a lot of time that you don't have to set up the whole Google SSO anymore. The only thing you need to add here is the credentials for Google and you're all set. That takes just a few minutes.

You can also see an application here called **Document Approval Application**. This is a complete application that has a workflow built in for approving documents and being able to upload documents. Of course, you can add to this template as well – even though it's a fully functional application, you can still add more functionality to it yourself. Maybe you would like this application to upload these documents to your own server once they are approved. That is something you can absolutely do with the platform.

There is even a template that has a guided building system built into it:

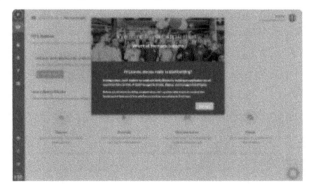

Betty Blocks: Task Manager

Create your own Task Manager
application following a step-by-step...

Figure 3.121 – The Guided building template

It will take you by the hand and guide you through making changes to an application. Every step is highlighted for you, so you won't get lost along the way. Of course, you have this book to help you with that, but if you want to get your first taste of a building Betty Blocks application, I'd recommend using that one. It's a great way to get started or just get a quick feel for the platform.

Summary

My Betty Blocks is all about the governance of your applications and users. It all starts with registering on the platform or being invited to an application or organization on the platform by an admin. Once you've completed your registration, you can get started on the platform.

Organizations hold your applications. Within an organization, you can have multiple applications and users. When a user is an admin of an organization, they can invite users to the applications of this organization and create new applications. These admins also have the option to create or merge sandbox applications.

We've talked about how you can create applications on the platform. You can create applications with or without using templates. Using a template can kickstart your development process. Once your applications have been created, you can invite users to your application. A user for an application can be an admin or a member. An admin can be built in the application, while a member can only access the back office of the application. This only allows them to modify data in the application through the back office.

I hope all of this gave you a good insight into what you can do with My Betty Blocks. Let's move on to actually using a template now and explore this template to see what you can actually build with Betty Blocks.

Questions

1. What are organizations used for?
2. What is a sandbox?
3. What are templates for?

Answers

1. Organizations are used to store all your applications in one central place. Here, it's easy for an admin to manage all the applications in your organization.
2. A sandbox allows you to create a copy of your production application so that you can make changes without breaking the production application. At a later stage, you can merge all these changes back into the production application.
3. Templates are pre-defined applications that allow you to get a quick start on your application development without having to reinvent the wheel every time.

4

An Introduction
to Data Modeling

In this chapter, we'll be talking about the data model in the Betty Blocks platform. If you have never done anything with data modeling or have only ever set up a database, this chapter will help you understand the basics of data modeling. We'll go over all the essential functionalities and build a small data model together to help you understand the concept better.

After creating the data model, we'll go into the back office of the platform. This part will help you bring your data model to life. It allows you to easily interact with, as well as access and insert data into, the data model.

We'll be covering the following topics in this chapter:

- Understanding what a data model is
- Creating your first model
- Relationships

After working through this chapter, you'll understand what a data model is, how to create a basic data model, some of the essential properties that go into a model and how you can create relationships between models, and of course, what the differences are between the types of relationships.

Understanding what a data model is

What is a data model? In short, a data model determines the structure of your data. The Betty Blocks platform consists of models, which have properties. Data models can also have relationships with each other. A data model is basically your database within the platform where you can store all the information your application needs or needs to store. So, what does this mean exactly if you have never used a data model or database before? Let's take a look at an example and try to present it in a way that you would normally see in a data model or database.

An example of a data model

Let's take a customer as an example. You want to create a data model that can hold different customers. You want to keep track of your customers and have them all in an app so you can quickly look them up and find their information or add more information to them. So, you would start by creating a Customer model in your data model. In a model, you can store information, in this case, information about customers. But you can't just store information here without telling the model what kind of information it can hold. This is what we call *properties* in Betty Blocks.

A property has a specific type, which helps it to store specific information. For example, the most common one is `Text (Single Line)`. This property is used to store short texts in your model. For example, the first name of a customer will be `Text (Single Line)`, which is also a city name can be or a postal code, for example. This text has a maximum length of 255 characters. This is a very common type in the world of data models and databases. There is also a property that can hold many more characters, so don't worry about not being able to store more characters. As the name already tells you, `Text (Single Line)` saves text on a single line, so it doesn't add a new line, for example. There are about 20 different properties in Betty Blocks. We'll have a look at the most common properties:

- `Text (Single Line)`: This one is mainly used to store short text with a maximum of 255 characters.

- `Text (Multiline)`: This one is mainly used to store long text that consists of multiple lines and can save new lines.

- `Date`: With this one, you can store dates in a pre-defined format.

- `Number with decimal`: With this one, you can store numbers with decimals.

- `Checkbox`: With this one, you can store `true` or `false` values. For example, if you need to `agree` to an agreement, you can check the checkbox.

- `Email`: This one allows you to store email addresses and applies checks to make sure a valid email address has been entered.

As you can see, there is also a specific one for email. Of course, you can also use `Text (Single Line)` for this, but the `Email` property has some specific features that a Text property doesn't have. Once you save an email address into the `Email` property, it will perform a check for you to see whether that email address is valid. This doesn't mean it checks whether it exists, but it checks whether the format of the email address is correct. For example, `john@doe.com` or `jane@doe.info` are valid email addresses as they have an `@` symbol splitting the name and the domain and a `.` splitting the domain and the extension. It also checks whether the extension is correct. If you enter `john@doe.c`, it will give you an error that the email address is incorrect and the data model won't allow you to save it. This is where the `Email` property helps you out, without having to set up this logic yourself. On the other hand, the `Text` property will just save any kind of email address, whether its syntax is correct or not, as long as it's shorter than 255 characters.

These are just six of the most common properties. There are many more properties that we will cover in *Chapter 7*. But, right now, I just want to give you an indication of what you can expect with models and properties.

So, the first model we want to create is a model to store information about our customers. The most logical place to start, then, is by naming the model. As a guideline, all model names in the platform are always singular and in English. This is not a rule, but it does make your life a little easier, especially, as you will see later, when it comes to the relationships between models. Of course, you are free to name them in any language you want. In our case, the most logical name for our customer model is `customer`. The next step that we need to think of is what information about our customers we want to store. In other words, which properties will we need to create in our customer model? Let's make a list:

- First name
- Last name
- Email address
- Date of birth
- Is Active
- Revenue last year

All these properties tell us something about the customer, for example, their first and last name. We could add a lot more properties here, but we'll keep it simple for now so it will be easier to remember. These six properties make up a good starting point for our customer model, but as you noticed, I haven't added the type of property behind these properties yet. Take a moment and have a look at the six property types described earlier, and try to assign the types you think would fit best on these six properties.

Alright, let's see whether your ideas match mine. In some cases, there is more than one type of property that might fit, but we'll take the ones that fit best for our use case:

- First name – `Text (Single Line)`
- Last name – `Text (Single Line)`
- Email address – `Email`
- Date of birth – `Date`
- Is Active – `Checkbox`
- Revenue last year – `Number with decimal`

The first two shouldn't be much of a surprise; a name is usually not longer than 255 characters, so it should fit perfectly in a `Text` property. The reason we split up first and last name is so you can work with both of them separately later on in your application. Of course, there are ways to split them later, but that will just make your work more difficult. By splitting them up now, you are making your life a lot easier. Always try to think of what the easiest way of saving the data is and how you would like to use it later in your application. This could save you a lot of time later on in the process of building your application.

Next, we have the email address as an `Email` property. I guess that doesn't surprise anyone either! Then, we have the date of birth saved as a `Date` property, which will help you add a date picker and even allow you to do simple and complex calculations with dates later on. The is active category is a `Checkbox` property because a checkbox is either true or false and nothing else. So, in this case, a customer will be active or inactive, based on the state of the checkbox. Checkboxes can make the logic simpler to implement later because you can simply check whether a checkbox is true or false. It can also help you easily filter your customers list to only see your active customers. Lastly, we have the revenue for last year in numbers with decimals. Since this is usually set in some kind of currency, this property would make the most sense. Of course, you can use `Text` as well, but since you usually want to do math with numbers, it's a lot easier to make the revenue last year category of the numbers with decimals type. You can make calculations with numbers, whereas with the `Text` type, you can't use the values in calculations even if they are numbers, because they have been saved as text and not numbers. This is something that's specific to programming and computers, so if you are not used to it yet, you will be after a little practice.

Creating your first model

Alright, now that we've thought about our first model, let's start creating our model in our Betty Blocks application. Let's start by creating a new application first:

1. Go to **My Betty Blocks** and then to your organization. Create a new application. Make sure you use the **Login** template for your application. You can name the application anything you would like. Once you've created your application, open it, and let's get started.

2. On the left, you will see the **builder bar.** When you open your application, you should be on the **Dashboard** page. Click on the purple highlighted data model icon (*Figure 4.1*) in the **builder bar** on the left side of your screen.

Figure 4.1 – The data model icon

Now, a side panel will open on the left of your screen. It should look like *Figure 4.2.*

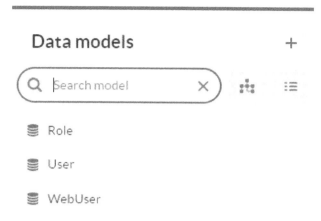

Figure 4.2 – The side panel for the Data models menu

It's a quick menu that allows you to quickly go to the models in your data model, but we want to see the schema view of our data model. For that, we'll need to click on the schema overview icon (*Figure 4.3*).

Figure 4.3 – The schema overview icon

3.   This should take you to the schema overview (*Figure 4.4*).

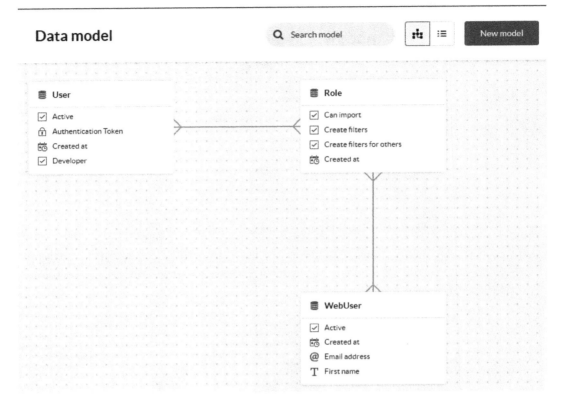

Figure 4.4 – The schema overview

As you can see, there are already some models present here. Every new Betty Blocks application comes with these three models already included. These three models can't be deleted, but they can be modified. We'll dive deeper into these in *Chapter 7*, right now we're going to focus on creating our own model first.

4.   To create a model, we need to click on the **New model** button in the top-right corner (*Figure 4.5*).

Figure 4.5 – The New model button

5. After clicking on the **New model** button, a dialog box will appear as shown in *Figure 4.6*.

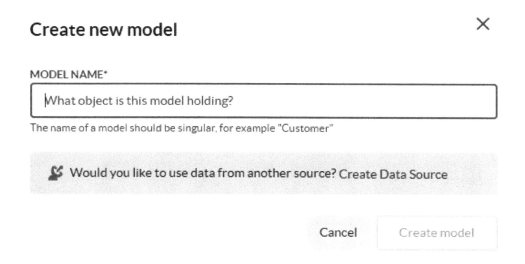

Figure 4.6 – The Create new model dialog box

As you can see in *Figure 4.6*, there is only one field we need to fill in here to create our model, which is the model name. Below the field, you can see a description that tells us that the model name should be singular, just like I mentioned a few pages back.

So, let's fill this field with the name of our `Customer` model, and then click on the **Create model** button. This should create a model for us.

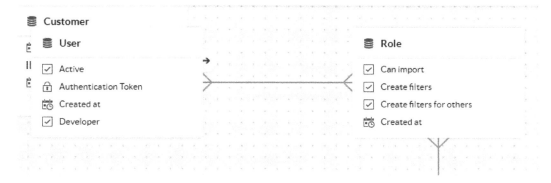

Figure 4.7 – The customer model hiding behind the User model

As you can see in *Figure 4.7*, our new customer model is hiding in the top-left corner behind the User model. It could also hide under another model if you can't see it. If you hover your mouse over the customer model on the top, we can drag it to another place on the canvas. When you have moved the model into an open position, it keeps the whole data model nice and clean by always putting your models in an orderly fashion, so it is easier to find them when needed.

Now that the model is easier to find, let's have a look at the model itself. As you might have noticed, it already has some properties added to it. By default, any model that you create will have the following properties (*Figure 4.8*):

- **Created at**
- **Id**
- **Updated at**

These three properties are mandatory for a model; you can't delete or modify these properties. But what do they do?

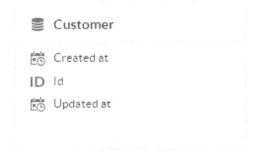

Figure 4.8 – The customer model on the canvas

Let's start with the Id. This property automatically generates a number, starting with 1, and increases the number by one every time you add a new record of data to this model. At this point, the data model is still empty, so there is no Id yet. The type of property for the Id is a serial, which is a unique property type that only the Id can have. You can't create this type yourself. At this point, there is no data in the model, but after we create the first record in our model (which we'll do in a later chapter on the *back office*), we'll see that the first record has an Id of 1.

Then, we've got the **Created at** and the **Updated at** properties. Both of these properties are of the date time type. These dates also get generated automatically by the platform once you create a record. They also can't be modified manually. The difference between the two is that **Created at** is only set once you create a record. It saves the date and the time when the record is created, so you will have a history of when someone created the record. As you might have guessed by now, **Updated at** is created at the same time, but it's also updated every time someone makes a change to that specific record. So, you can track when the last change was made to that record.

All three of these properties are important for most applications, so they are never in the way. Even if they might be, you can always choose to ignore them and not let them show up in your actual application later. Let's continue by adding some properties of our own to our model.

When you hover over the word **Customer** in your customer data model, it should show that the **Customer** has been underlined. Click on **Customer**, and it should open the details of your model (see *Figure 4.9*).

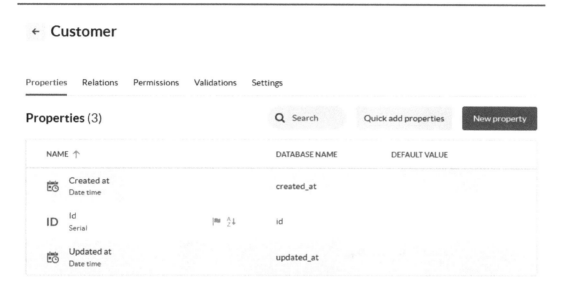

Figure 4.9 – The Customer model details

Here, you can add new properties to your model. There are two ways of doing this. You can quickly add multiple properties at once or you can add properties one by one. So, when do you use which one? The **Quick add** function allows you to quickly add properties by just pressing keys on your keyboard and not worrying about any other options at that point. This means you can quickly type the name of your property, choose the type you want, and go on to the next one. This usually works very well, especially if you have a lot of properties to add. The **New property** button is for when you want to add a new property while selecting specific options for these properties. In this chapter, we'll use the **New property** button and we'll discuss three of the most commonly used options for your properties.

So, let's click on the **New property** button and start adding our first property. A side panel should open and it should look like *Figure 4.10*.

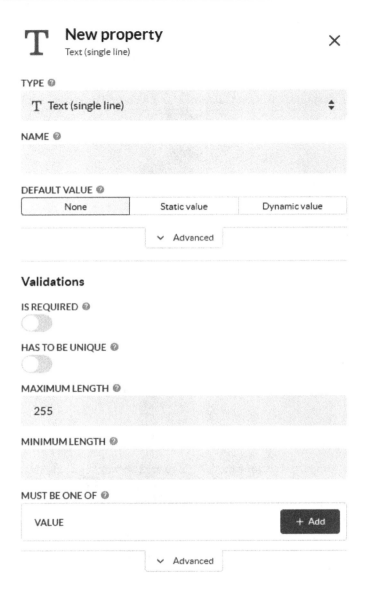

Figure 4.10 – The New property panel

The first thing we see is the type of property. If you open this drop-down menu, you'll see a lot of different properties, a lot of which we haven't covered yet. We'll go over these in a later chapter, so you won't be overwhelmed by them now. Right now, for our first property, we don't need to change the type, so let's leave it as **Text (single line)**.

Now, we need to name our new property. You can name your property almost anything. There are a few reserved words in the platform, if you type one of those words you'll be notified by the platform and you'll need to change it before you can save it. In our example, this won't happen. If it does happen, just choose a name that is similar to your first chosen word. In this case, let's name our property `First name`.

Next is the **DEFAULT VALUE** option. This option is present on most other properties as well. As you can see, next to each option there is a question mark. This explains what each field does.

Default values are preset values defined for a property. So, we could set it to be `Joe`, for example, so that it will always be `Joe`. But that wouldn't make any sense for a first name. So, by default, it's set to **None** so it won't use any default values. But if you want to set a static value, you can choose that option. There is also a dynamic value available, which can create a value based on some logic. For now, we'll go with **None**.

Next up are the validations. I will explain the two options to you. The **IS REQUIRED** option makes sure that the user who wants to input a record into the customer model sets the first name value. If they don't, they'll get an error telling them that the first name is required.

The **HAS TO BE UNIQUE** option checks whether the value for the `First name` property is unique across the entire customer model. So, if you were to switch this one on too, it would mean that you can only have one person with the name `Joe` in your customer model.

Let's switch the **IS REQUIRED** option to on so that including the first name is mandatory. We'll go over the other options in *Chapter 7*. See *Figure 4.11* for an example of how your new property should look now in the panel. Then hit the **Add** button at the bottom of the panel.

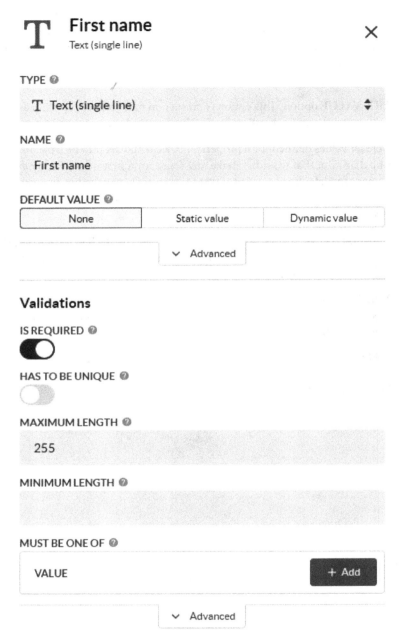

Figure 4.11 – The First name property set up

You'll notice that as soon as you hit the **Save** button, the First name property is added to the properties on the left (*Figure 4.12*).

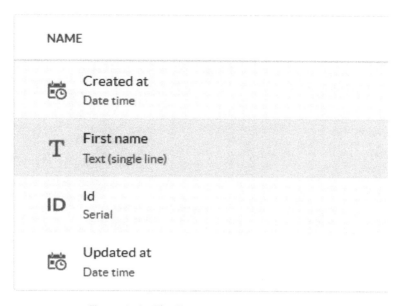

Properties (4)

NAME
📅 **Created at**   Date time
T **First name**   Text (single line)
ID **Id**   Serial
📅 **Updated at**   Date time

Figure 4.12 – The First name property added

Now, click on the **New property** button again. It should allow you to create a new property. The fields should be empty, and you are ready to create the next property. Add the following properties to your customer model with the following options:

- `Last name` – **Text (Single Line)** – **IS REQUIRED**
- `Email address` – **Email** – **IS REQUIRED, HAS TO BE UNIQUE**
- `Date of birth` – **Date** – **IS REQUIRED**
- `Is Active` – **Checkbox**
- `Revenue last year` – **Number with decimal**

Once you added all of your properties, the property overview should look like *Figure 4.13*.

NAME			DATABASE NAME	DEFAULT VALUE
Created at Date time			created_at	
Date of birth Date			date_of_birth	
First name Text (single line)			first_name	
Id Serial			id	
Is active Checkbox			is_active	✓
Last name Text (single line)			last_name	
Revenue lasts year Number with decimal			revenue_lasts_year	
Updated at Date time			updated_at	

Figure 4.13 – The property overview with all properties added

If you would like to zoom in on one of the properties, for example, the **Date of Birth** (because here you can set up the date format as well and date formats differ throughout the world), let's quickly go over how you can change it.

The default date format is the standard Dutch format (Betty Blocks was developed there, so you can't blame us). But, of course, you can change it to suit your needs. As you can see now, it's set to DD-MM-YYYY. You can change the dashes to slashes, for example, so you'll get DD/MM/YYYY, or you can change it to the US format by changing the DD and MM around. But there are a lot more options available as well; you can even add the day of the week to it or the week number itself. All of these options are described in the documentation of Betty Blocks. It would take a chapter to talk about that here, and since it's something that you'll do once when it's needed, it's easier just to quickly look it up when changing it than to learn all of it. To quickly go to the documentation, click on the **Learn how formatting works** link under the FORMAT option (*Figure 4.14*).

FORMAT ❔

DD-MM-YYYY

Example outcome: 24-03-2022   Learn how formatting works

Figure 4.14 – Date formatting

Let's go back to the schema overview of our data model by clicking on the arrow in the upper left corner (*Figure 4.15*). As you can see now, all the properties are showing in our model on the schema view as well. Before I show you how to start adding some data to your model, let's first dive into the relationships between models in the next section.

Figure 4.15 – The back arrow

Relationships

Right now, we've created one new model: Customer. But we want to add more to our data model. In our case, we would like to associate an address with each customer. There are different ways of doing this. We can, of course, add a couple of new properties to our customer model. This would mean that you can register an address on the customer. There is nothing wrong with this at all, but let's say you would like to add multiple addresses for your customer. Your customer may, for example, have a home address, but also a work address. This would mean that you would have to add the same properties twice with different names. This could get messy really quickly if you want to add more than a few addresses for a customer. So, in this case, we will add another model named Address to our data model. Let's click on the new button in the top-right corner of the model schema overview. Let's name our new model Address and save it. Most likely, it will be in the top-left corner of your schema overview. Drag it near the customer model.

Figure 4.16 – The Address model added

Now, it's time to add some properties to our Address model. Click on the Address name in the Address model to open it up. Click on the **New property** button and let's add the following properties with their options:

- Street – **Text (single line)** – **IS REQUIRED**

- House number – **Text (single line)** – **IS REQUIRED**

- Postcode – **Text (single line)**

- City – **Text (single line)** – **IS REQUIRED**

After you've added these properties, we're going to add one more property, a new type of property that we haven't used before. This is the List property. This property allows users to select a value from a list of values. These values will already be determined at the creation of the property itself. In our case, it's going to be a list of address kinds (the reason we are not using the word type here is because this is a reserved name in the platform).

So, let's add another property of the List type and name it Kind. As you can see under the name, there is a button called **Add value**; click on it to add a new value to the list and name it House. Then, add two more, Work and Post, so we'll have three in total. Also, set **IS REQUIRED** on. It should look like *Figure 4.17*.

Figure 4.17 – The List property for Kind

So, this list will allow the user, later on, to choose between three kinds of addresses namely: House, Work, and Post. This will make it a lot easier to identify which address you'll be selecting, but we're not done yet as we haven't created any relationship between the two models. So, why would you like to have a relationship between your models? This could be for a number of reasons, but the most important reason is that the data is related to each other in some way. In this case, the relationship between the two models is that the customer can have one or more addresses. This also works from the addresses' perspective as an address belongs to a customer. Because that specific address belongs only to that specific customer, it can only have one customer connected to it. So, in the Betty Blocks platform, there are three kinds of relationships:

- **Belongs to**
- **Has many**
- **Has and belongs to many**

Each of these has a specific function. Right now, we will just focus on the first two. Let's jump into an example to make this a little easier. If you have an order system, for example, this system will record orders for your company. You'll need a model called `Order`, which will hold all of these orders. However, an order will need a customer as well, so you know where the order has to go. So, in this case, the relationship between the order and the customer will be a **Belongs to** relationship. This is because the order belongs to a customer. The **Belongs to** relationship tells you what kind of relationship it is without needing any further details. So, what does this mean for your application? It means that for each order, you can have one customer, which is the customer that has placed that order. The customer model will be filled with all the customers, where on the order you can pick a customer from, because of the relationship. This also means that when you have three orders in your system, they could all belong to one customer. This is the reason why you may want to create a customer model and not put the customer properties on the order because then you will have to type in the customer manually every time. Now, you can select it from the order and connect it to your orders. This minimizes the chance of mistakes. Now, the other relationship also comes into play. This other relationship is the **Has many** relationship. If you look at this from the customer's perspective, a customer can have many orders – they could have 0, 1, or even 1,000 orders. So, when you create a relationship from the customer to the order and you want to register multiple orders on your customers, you select the **Has many** relationship type. So, basically, when you select a **Belongs to** relationship, the other side of that relationship automatically becomes a **Has many** relationship, while if you do it the other way around, it becomes **Belongs to**. I hope this makes sense, but if it doesn't, don't worry about it too much right now. We'll create the actual relationship now, then we'll go to the *back office* where you can see it all in action. Hopefully, it will make more sense then.

Click on the **Back** button in the top-left corner to go back to the model schema overview. Let's create our relationship between the `Customer` and `Address` models. In order to do this, we'll need to drag an arrow from one model to another. You can do this by hovering over the customer model, for example, and an arrow should appear (see *Figure 4.18*).

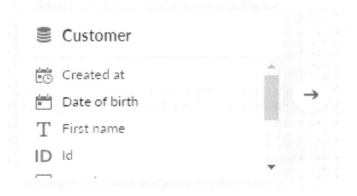

Figure 4.18 – The relationship arrow on the right

If you click on the arrow and hold your mouse button, the arrow should change into a line. This should allow you to drag the arrow in any direction (*Figure 4.18*). In our case, we want to connect to the `Address` model. So, drag it onto the `Address` model and release your mouse button.

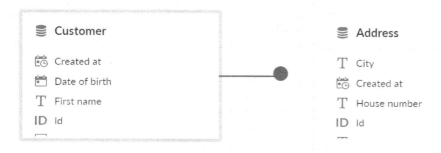

Figure 4.19 – Dragging the arrow into a line

After you release your mouse button, a dialog box should appear where you can select the relationship type (*Figure 4.19*). As you can see here, the three relationship types that we've talked about before are shown here. Each type even has a description based on your models:

- **Belongs to** – Many customers belong to one address
- **Has many** – One customer has many addresses
- **Has and belongs to many** – Many customers have and belong to many addresses

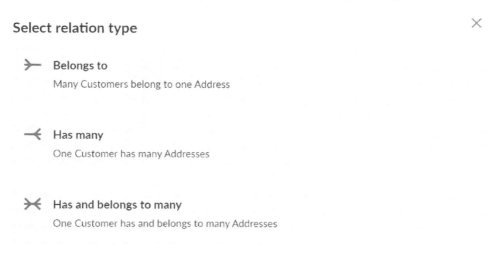

Figure 4.20 – The relationship type choice

This should make it a lot easier to make the right decision on which relationship you would like to create between the two models. So, let's think about what kind of relationship we would like to create between our models again. Our customers should have the possibility to have multiple addresses registered. So, if we look at the options we have then **Has many** is the best to use here as one customer has many addresses. So, let's select **Has many** here. Now, our model schema should look like *Figure 4.21*.

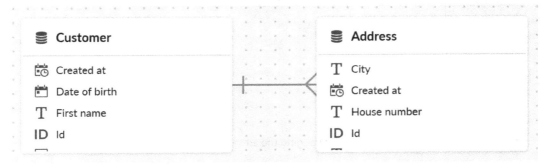

Figure 4.21 – The Customer and Address models with a Has many relationship

Now, we have our small data model, but we can't enter any data yet because that's done on a page that you'll build in the page builder or the back office. We'll get to that in the next chapter where we will discuss the back office.

Another relationship example

In order to help you better understand the basics of relationships between models, let's go over a few more examples.

Earlier, we spoke about the `Order` model, which has a relationship with a customer. In that example, the order always had one customer. In the data model, that would be a **Belongs to** relationship, but it also meant that the customer could have more than one order, which would make the other relationship a **Has many** relationship. But an order will also have order lines. An order line is the line on an order that describes what a person has ordered, for example, "10 boxes of paper for 10 euros." There could be 1 or 100 order lines, for example, since every customer can add something completely different. In this case, an order can have many order lines, while an order line always belongs to one order. After this, your data model will look like *Figure 4.22*.

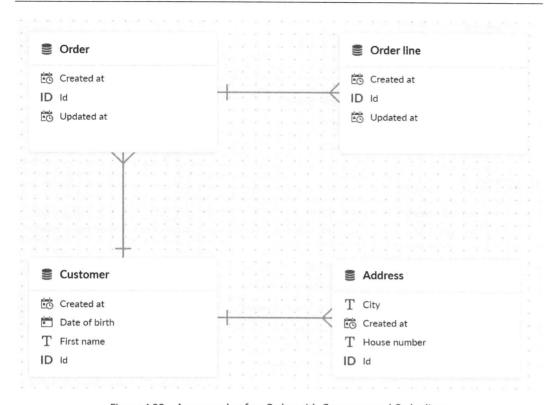

Figure 4.22 – An example of an Order with Customer and Order lines

Well, this concludes the relationship example. Let's summarize what we've learned in this chapter.

Summary

This chapter was all about the data model basics. We created our first models in a data model and took a look at the properties of a model and the relationships between models.

A data model is a collection of models that represent the structure of the data that you would like to save in your application. Models contain properties that define exactly what data is stored in your models. Each property has a specific type that ensures that specific data types are being stored in the right property. For example, if you want to store a date, you can use a data property. The data property will make sure the format of the date is correct and also enables you to later do calculations with your dates in an easy way.

Then, we looked at the relationships between models. Relationships enable us to store data in an effective way so that a customer, for example, can have multiple addresses without having to create a lot of properties for different addresses. In the next chapter, we'll introduce you to the page builder.

Questions

1.   What is a data model?

2.   What is a model?

3.   What are the most common property types?

4.   What's the best way to name a model?

5.   What types of relationships are there?

Answers

1.   A data model determines the structure of your data. In the platform, it consists of models, which have properties. They can also have relationships with each other. It's the place where all of your data is stored within the platform.

2.   A model is a place where you store all data that is directly related to each other, for example, a model could be a customer, an address, or an invoice

3.   The most common property types are `Text`, `Number`, `List`, `Date`, and `Checkbox`.

4.   A model name should always represent the data that it is going to be holding. The easiest way to do this is to name it in English, but it can also be done in any other language.

5.   Betty Blocks has three relationship types: **Belongs to**, **Has many**, and **Has and belongs to many**.

5
The Page Builder

In this chapter, we'll discuss the page builder in the Betty Blocks platform. We'll go over what exactly the page builder is and what you can do with it. We'll create our first page by using a template and add some components to our page with the template to explore the components. And lastly, we'll discuss how you can create a simple prototype by using a page template, so you can showcase your ideas by giving them form through the page builder.

These are the main sections to be covered in this chapter:

- What is the page builder?

- Our first page, based on a page template

- Modifying a page template

- Creating a prototype page

After this chapter, you should have a basic knowledge of what kind of pages you can build with the page builder and an understanding of the first set of components in the page builder, which will help you to build your first pages.

We'll also explore the page templates, which can help you quickly build a page with minimal effort. We'll discuss the different types of pages that are currently available. And to practice some more with some of the components, we'll make some modifications to the page template. Lastly, we'll create a page from scratch for a prototype, so you can learn how to build a page that can help you get your ideas across without having to build any actual functionality behind the page yet.

Let's get started by discussing what the page builder actually is.

What is the page builder?

The page builder is a drag-and-drop editor used to create web pages. It allows you to do most of the things with it that you are allowed to do with code, just in a much simpler way because you don't have to remember all the code for creating a page. It also allows you to integrate with the data model that you've created so you can display all the data on your application's pages. Backend logic is handled by the actions part of the Betty Blocks platform. We haven't touched this part yet, but it basically allows you to create backend logic with drag and drop as well. This will be covered in another chapter.

With most no-code tools, you can get stuck because they don't support something that you might want to do. In the Betty Blocks platform, you can add functionality and extend the functionality provided by the platform by using code. This doesn't mean that someone who can write code needs to create this part by adding code to your page; instead, they can create a no-code block using code to enable other developers to use it. This allows the citizen developer or no-code developer to build the application they want without making sacrifices in functionality. And this new functionality can be shared across different applications as well, so it becomes part of their whole Betty Blocks environment.

For this chapter, we will use the application that we created in the previous chapter.

So, let's get back to the page builder. Let's explore some of the basics first since most of what I just mentioned is meant for when we can actually build an application.

Opening the page builder

In the builder menu on the left side of your screen, you'll see a little white paper icon, which, when you hover over it, should turn blue, just like the following screenshot:

Figure 5.1 – The page builder in the builder bar

Click on this paper icon, and the quick view of the **Page overview** page should open. This quick overview is great for quickly opening the page you are looking for or quickly creating a new page, but in this case, we want to go to the **Page overview** page. So, we need to click on the overview button. This is the gray button with a list view icon next to the search bar, and you can see an example of this button in *Figure 5.2*:

Figure 5.2 – The page overview button

On the **Page overview** page, you get an overview of all the pages you've created so far. In our case, this page is empty since we haven't created any pages yet. There are three categories of pages, as you can see at the top: **All**, **Public pages**, and **Authenticated pages**. These three types just differentiate your pages, so it's easier to find them. **Public pages** are pages that anyone can see without being logged in, while **Authenticated pages** require users to be logged in first before they can access the page. And of course, on the **All** page, you can find both types of pages.

There is also an option to search for your page based on its name. This makes it easier to find your page once you start creating more pages. And then, there is the **New page** button, which allows you to create a new page. So, let's start there next.

Our first page, based on the page template

Let's click on the **New page** button in the top-right corner. This should open the **Page creation** page (see *Figure 5.3*). On this page, you'll get an overview of all the public page templates that Betty Blocks offers out of the box. The **Start from scratch** button allows you to create a blank new page if none of the templates are what you are looking for.

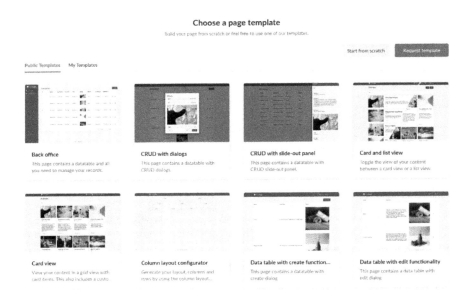

Figure 5.3 – The Page creation page

There is also an option to create your own page templates, which you can find in the **My Templates** tab. Right now, these templates are probably empty. There is also a button in the top-right corner for requesting templates. You can use the **Request template** button to request a specific template that you think is a good addition to the platform.

But we are here to explore our first page, so let's use a template to do so. This should help you understand some of the basics of the page builder without having to build a whole page yourself. Scroll down until you see the **Inspirational Dashboard** template; this template has no data on it from a data model. First, we want to explore the page builder with a static page to help you understand it better, which is why we are going to use the **Inspirational dashboard** template. Hover over the **Inspirational dashboard** template with your mouse and click on the **Use** button. This should open up a dialog box that asks you to name the page and set the path of the page (*Figure 5.4*).

The name of the page can be anything you want it to be; of course, it would better to give it a descriptive name, so you can find it easily later. The **PAGE PATH** field is the name of the link your page will get when you want to visit it through your browser. By default, this is going to be derived from the value you enter in the **PAGE NAME** field. If your page name is very long, you can make changes to the path to keep it simple and more understandable in the browser. Let's name our page Example, and let's leave the path as the default value.

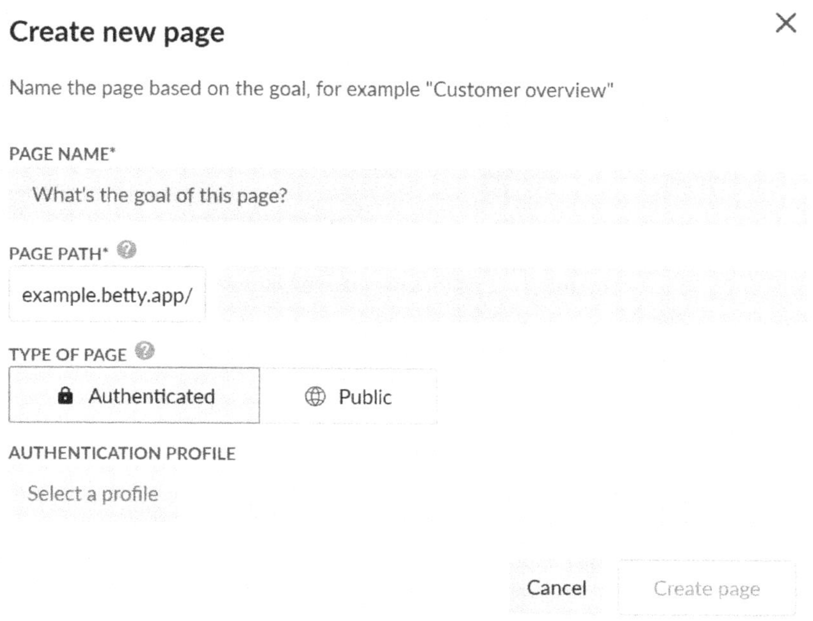

Create new page   ✕

Name the page based on the goal, for example "Customer overview"

PAGE NAME*

What's the goal of this page?

PAGE PATH* ⦾

example.betty.app/

TYPE OF PAGE ⦾

| 🔒 Authenticated | ⊕ Public |

AUTHENTICATION PROFILE

Select a profile

Cancel   Create page

Figure 5.4 – The create new page dialog

The third option that we see is the **TYPE OF PAGE** option, which decides whether the page is going to be publicly available or behind an authentication. Since we haven't set up any authentication yet, let's choose **Public**. You'll see that the **AUTHENTICATION PROFILE** option disappears, so we don't have to decide this in this case. Let's hit the **Create page** button, which redirects us to the second part of the creation of this page (*Figure 5.5*). For now, we'll skip this part, as we'll discuss this in more detail in the *ToDo Application* chapter. Let's click on **Add without configuration** to skip this section. Our page should be created for us now.

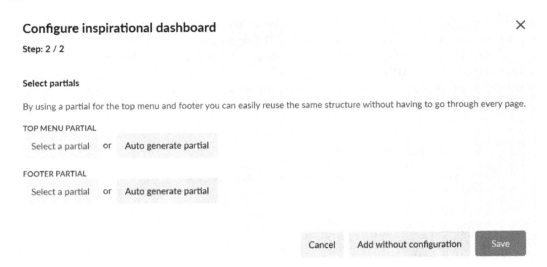

Figure 5.5 – Step 2 of the page creation

Your page should now look something like the screenshot shown in *Figure 5.6*. You've now arrived at the page builder.

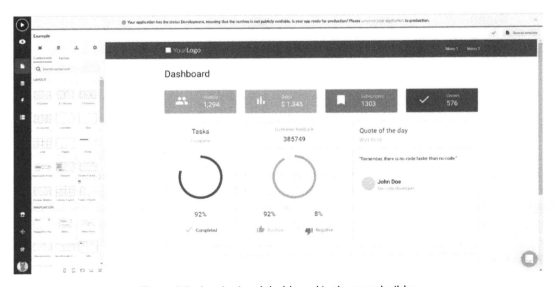

Figure 5.6 – Inspirational dashboard in the page builder

So, what exactly can we see here? Because, right now, this might be a bit overwhelming. Let's go over some of the basics of the page builder first. We'll go step by step, so you'll know how to use all parts of the platform.

The page builder side menu

Let's start on the left side of the screen, next to the builder bar. In *Figure 5.7*, we see the tabs that make up the menu of the page builder, together with the components under them.

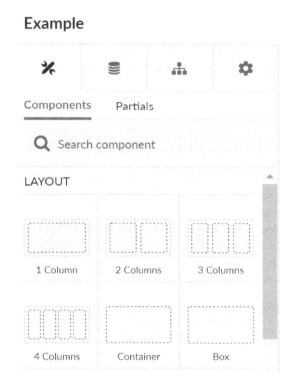

Figure 5.7 – Page builder tabs and components

The four tabs are as follows:

- **Components**
- **Variables**
- **Component tree**
- **Page settings**

In the **Components** tab, you will find all the components that you can use to create your pages. You can drag a component from here onto the canvas and drop it in the place where you would like it to appear.

The **Variables** tab allows you to set up variables that can help you to send over data from one page to another. I understand that right now, this might not make a lot of sense. We won't go into this right now; just be aware that it exists. We'll get back to this in *Chapter 8, ToDo Application*.

The **Component tree** tab is very important because you'll be using a lot of different components on your page, which may overlap each other. This could mean that some components might be harder to select later on. The component tree allows you to see all components on your page in a nice tree hierarchical overview and select them from there, so you don't have to visually search through your page.

The **Page settings** tab allows you to make changes to the name of your page and the path that we set up earlier when we were creating the page. Among other things, they are more technical.

Components

Components are the basic elements that make up a page in the page builder. They can be small and simple, such as a column. For example, a column is a component that you can use to create the layout of your page by adding different columns and sizing them according to your design. Components can also be more extensive, such as a dialog, which consists of more than one component essentially. It's like a pop-up window where you can inform your users by showing some info. The difference between these two examples is that the dialog is basically a set of different, smaller components that makes up the dialog. For example, with the dialog, you have the dialog itself but also a header inside it and a few buttons to close the dialog by default, whereas the column is exactly what it says it is, a column that allows you to set up the basic layout of the page. But we're getting into a lot of detail already. Let's focus first on what types of components there are.

There are a few types of components that make up the page builder. They are divided into the following different sections:

- **Layout**
- **Navigation**
- **Content**
- **Button**
- **Data**
- **Form**
- **List**
- **Card**
- **Logic**

Each section is there to make it easier to find a specific component that you might need on your page. Of course, there is a search option as well that can help you find the correct component.

You can drag and drop each of these components, and the part where you drop the components is what is called the canvas. In our example right now, we already have components on our canvas. We can add more by grabbing one from the components section and dragging it onto the canvas, but there might be some components that can only be dropped in a specific component. For example, most components expect at least a column to be on the page first before they can be dropped on the page. An empty page can only have layout components first before you can, for example, drop text onto the page. We'll explore the different options of which components to drop first in the next section of this chapter.

Now, let's take a look at the components that are currently on your canvas. If you click on the **YourLogo** image at the top of the canvas, you can select the **AppBar** component (*Figure 5.8*):

Figure 5.8 – The AppBar component selected

You'll see that the selected component has a purple box around it, indicating the area of the **AppBar** component on the canvas. You'll also notice that on the left side of your screen, the components are gone and have been replaced with the options for this specific component (*Figure 5.9*):

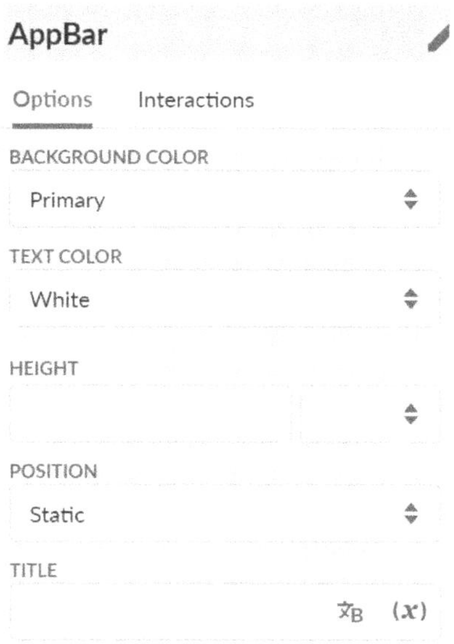

AppBar

Options   Interactions

BACKGROUND COLOR

Primary

TEXT COLOR

White

HEIGHT

POSITION

Static

TITLE

$\bar{x}_B$ (x)

Figure 5.9 – The AppBar options

These options allow you to make modifications to this component so that it suits your needs. You can change colors, change the image, adjust the height, and many other things. We'll not be going over each component's options in this book because that would take too much time. Most of these options speak for themselves, but in the different chapters, we will go over the most important options. For now, it's important to know where to find the options for your components. So, if you click on a component, it will show you its options on the left side of your screen.

Let's click on the **TITLE** dashboard on the canvas to see what options will appear for a title. As you can see, other options appear for the title, such as the option to set the title itself. You can change the type of title you want in order to make it smaller or larger, and there are options for the alignment of your text as well. If you have some familiarity with the basics of the web, most of these options should be easy to understand.

The component tree

Now, let's have a look at what the structure of the page looks like. To do this, we'll need to open the **Component tree** tab. It's located on the left side of the screen and is the third tab (*Figure 5.10*).

Figure 5.10 – The component tree tab selected

The component tree shows you the structure of a page on the canvas by showing you all the components on it in a hierarchical structure. Since a lot of components live in other components, it's easy to get lost or not be able to select or find the component you might want to edit. This is where the component tree comes in handy. It allows you to see all the components and easily select them (*Figure 5.11*).

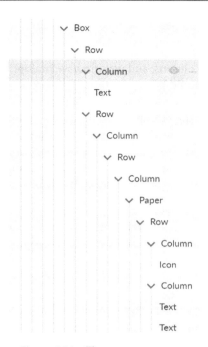

Figure 5.11 – The component tree

In the **Component tree** tab, as shown in *Figure 5.11*, you can see a part that represents the dashboard title until the first orange visitors' box. If you hover over the components in the component tree, you'll see all the components highlighted in the canvas, so you'll know exactly where they are located. In *Figure 5.12*, we are hovering over the column component in the component tree, while on the canvas, you can see it highlighted.

Figure 5.12 – Highlighted column

If you look at the visitor's box, it looks like one simple box, but when you look at the component tree, you can see it consists of multiple components. Let's break this down to get a better understanding.

A row helps you to order all your columns, and you can have multiple columns in a row. And when set up properly, those columns should be on one line. In this case, each of these boxes, **Visitors**, **Sales**, **Subscribers**, and **Orders**, has its own column, so they are separated easily. Then we use **Paper**, which creates an effect that makes the box appear to be lifted a bit on the canvas. Inside **Paper**, we have another **Row** because want to show the icon and the **Visitors** with the numbers on each side. A row with two columns makes it easy to split those evenly. Then, inside these columns, we have an icon and the other two **Text** components. It sounds like a lot, but once you get into creating your own pages, this will all make sense very quickly. I hope this gives you some idea of how to organize components on your page.

In the next section, let's add some components of our own to the canvas and modify them to get a better feeling of how you can drag and drop to create your own page.

Creating a layout

In this section, we're going to create our page layout to get a feeling of the drag-and-drop functionality and explain some of the layout components and their options.

Before we get started, let's create a new page, so we have a blank page to get started with. Click on the page builder icon in the builder bar on the left side and when the sliding pane opens, click on the blue cross button in the top-right corner. This should bring you to the **Choose a template** page, and here we can create a new page based on a template or create one from scratch. We'll start from scratch now to play around with some of the layout components. You'll see a gray **Start from scratch** button at the top (*Figure 5.13*). Press it and name the page Layout Test, set it to **Public**, and click on **Create a page**. You should see an empty canvas now in the page builder.

Figure 5.13 – The Start from scratch and Request template buttons

The most important component in the page builder is the **Column** component. This component is used for a few important things:

- It's the primary component used to be able to drop other components into
- It allows you to create your page layout
- It allows you to set up the different page layouts for desktop, tablet, and mobile

Why is it the primary component used to drop other components into? Because without the **Layout** component first on your canvas, you can't drop any other component on the page. Give it a try by grabbing a text component and dragging it onto the canvas. You'll see that nothing happens because it can only be dropped into other components, such as the column component. A column is based on the 12-column principle. The reason why it's 12 is that 12 can be divided by 1, 2, 3, 4, 6, and 12, so it is very flexible.

Our first columns

You can create a layout with the columns, and, as you might have noticed, there are four different column components. Each component has a specific number of columns, from one to four. These aren't the only possibilities, but they are the most common, so that's why they are offered by default. We'll explore how to change them to your liking in this section. For starters, let's drag **1 Column** onto our canvas. Click on the component and hold your mouse button and release it once you are over the canvas. Your page should now look like *Figure 5.14*. A single column is displayed at the top of your canvas.

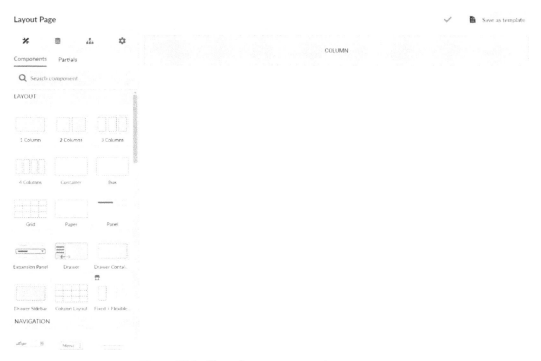

Figure 5.14 – The column component on your canvas

Let's also drag a **2 Columns** component under our first **Column** component onto the canvas. Now that we have two columns, we have a little bit of a layout but this layout won't be visible on the page, since the gray part in the column is just a placeholder so that you know where you can put your other components in an empty column. In our top column let's place some text, so we can see what the column will look like with another component placed inside. Search for the **Text** component in your component library on the left (*Figure 5.15*) and drag it into the top **Column** component.

Figure 5.15 – Searching for the Text component

Your page should now look like *Figure 5.16*.

Figure 5.16 – Column with text and two empty columns

Now click on the **Text** component on your page. The left side of your screen should now display the **Text** component options. Type some random text in the **CONTENT** box so we have some text in the column. As you can see, it will appear immediately after entering the text in the **CONTENT** box (*Figure 5.17*). As you can see, the gray placeholder area is now gone, and since it now has some text inside, it shows the true column. Officially a column has no height or border, so on the page itself, it will just appear as white space without any other component inside it or without changing the options of the column to show a color or give it a specific height.

Figure 5.17 – Text added to the first column

To see what happens when you change the height of a column, let's change the background color and the height of the two columns that are at the bottom of the page right now. Select the first column and on the left-side menu, click on **Styling**; now you should see **Height** and **Background color**. We can use these settings to change the height and the background color. For the first column, let's choose a height of 100 pixels. Enter 100 in the height box and change the second dropdown to **px**. Then set the **Background color** field to **success**, and you should see that the outer part of the column changes to green. The gray part will also be green, but since we don't have any other component inside the column, it will still show us the placeholder. When we go to the live page, we'll see that it is completely green.

Let's do the same for the second column, but set it to a height of 50 pixels and set the **Background color** field to **danger**, so it turns red. Now, in order to see our page live, we need to hit the play button in the builder menu. You can find it at the top of the builder menu on the left. Your page should look like *Figure 5.18*.

Figure 5.18 – The live page

As you can see, the gray placeholders are gone, and the columns show as green and red bars now. You can play around a little by changing the height and the background color and maybe placing a text component inside the colored components to practice.

Desktop, tablet, and mobile mode

The page builder has modes for desktop, tablet, and mobile views. This should help you to develop more quickly for each of these modes. If you don't want to develop one of them, you don't have to do anything. Just turn on the mode you want to develop for and make sure it looks good in that specific mode. You might be wondering, *How do I change the page builder to that specific mode?* At the time of writing this book, the options for changing the layout in the page builder were in the bottom left corner of the page builder, but this might have changed to be on the top center of the page builder. The different modes are shown in *Figure 5.19*.

Figure 5.19 – The different modes in the page builder

From left to right, we see the phone, tablet vertical, tablet horizontal, laptop, and full-screen mode. By default, it's set to full screen. Try changing the mode, and you'll see that the screen turns into a phone or tablet, for example.

Let's see how we can change our current page, so it works well on a desktop and on a phone. We've got the text and the two colored columns. We'll use the phone and the full-screen option for this. When you change it to the phone option, your screen will look something like *Figure 5.20*.

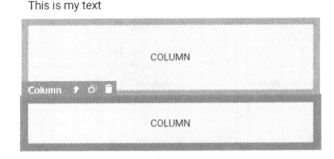

Figure 5.20 – Our page on a phone screen

As you can see, it automatically changes to look better on a phone. So, out of the box, it's already taking care of your phone screen, but what if this was not the case? If you click on one of the columns and look at the options that appear on the left side of the screen, you'll see four options for the column width so that for each type of device, you can set it up differently. A column in the page builder is always measured from 1 to 12. Why is this 12? That's because 12 allows you to easily divide your screen. For example, if you want 4 identical columns, you set them to 3, and since 4 * 3 = 12, they fill the whole width of the screen. This is something that a lot of frameworks use for websites, so this standard is also used in Betty Blocks.

So, let's make a change to our columns so that we can see it in action. Stay in mobile mode and change the options on the left for the mobile to 6 for both columns (left and right).

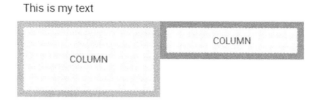

Figure 5.21 – Our columns with a width of 6 in mobile mode

As you can see, this wouldn't support much text on a mobile phone. So, the 12 on the phone would be much better. Let's change it back to 12 for mobile. Let's drag in a **4 Columns** component from the **Components** tab (*Figure 5.22*).

4 Columns

Figure 5.22 – The 4 Columns component

As you can see, in mobile mode, it is automatically set to a width of 6, while in desktop mode, it's set to a width of 3. If we change the mode to full screen, you can see that it will span the whole screen with four columns. If you select one of the columns and look at the column width options on the left side, you'll see there are fifteen different options there to choose from. We won't go through all of them, but you can play around with these options a little on your own once you've finished this chapter to get a feel for how to set up your columns.

Columns in columns

You can also drag another column into an existing column, this will allow you to make even more precise layouts. Let's drag a **2 Columns** component onto our canvas in full-screen mode. And in the first column of this **2 Columns** component, let's drag another **2 Columns** component. Your page should now look like *Figure 5.23*.

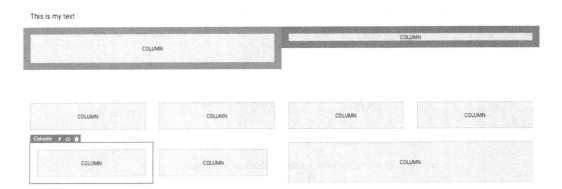

Figure 5.23 – Your page set up with different columns

Now you have two columns and one large column next to it. Of course, there are many other ways to achieve this, but for now, this should give you an idea of how to work with columns and set up your page layout.

Summary

In this chapter, we started using the page builder for the first time. The page builder helps you to create pages by dragging and dropping components onto a canvas. You learned how to get to the page builder by selecting the page icon from the builder bar and then selecting the overview icon. This took us to the page builder overview page. Next, we created our first page based on a template, which taught you how to create a page, how to name your page, and what options you can use while creating a page from a template. We then went through some of the basics of the page builder and the components in this template to get a feeling of the page builder. We touched on the components, the component tree, and the options of a component.

Then we created a new page that was completely empty. Here, we learned about creating layouts with a column. We used different options in columns to set up a page and also learned how we can change our view mode from full screen to mobile or tablet so you can also develop mobile or tablet pages.

In the next chapter, we'll create our first page as a prototype, which can be used to demo your idea for an application before you start building it.

Questions

1. What can you do with the page builder?
2. What is a page template?
3. What is the component tree?
4. Can you also build mobile sites?

Answers

1. With the page builder, you can build the frontend part of your application. It allows you to drag and drop your page together and style it from there.

2. A page template is a predefined page that already has the page set up for you. It also allows you to add data to it from a data model and make other configurations, so you don't have to manually set up a lot of the page.

3. The component tree allows you to quickly navigate through the components on your page. Some components might be hidden on your page because other components are overlapping them. With the component tree, you can easily access them.

4. With the page builder, you can build websites for desktops, mobiles, and tablets.

Creating a New Application from a Template

In this chapter, we are going to create a new application using a template. We will show you what a template looks like and how it works in this chapter. After we've created this application, we'll take you through the development environment and show you the basics of the page builder, actions, and data model. After that, we will go through the actual application that you've created to see how it actually works based on everything you've seen in the development sandbox.

We'll cover the following main topics in this chapter:

- Creating an application from a template

- Entering the development environment

- Launching the application

By the end of this chapter, you will know how to create an application from a template, which will allow you to quickly set up an application with a lot of functionality already in place. We'll take a tour of the development environment and show you some of the basic functionalities, and lastly, we'll launch the actual application so that we can explore it in depth.

Creating the application from a template

We will give ourselves a head start by creating an application using a template. In our case, we will be using the *Questions and Answers* application template. This application will give you good experience with the platform and its features.

Open the **My Betty Blocks** environment and navigate to your organization. This page contains information about your organization and the applications that you and/or others from your team/organization have created.

If the organization has just been created, there will be no applications on this page.

We will create a new application by clicking on the **Create application** button, as shown in *Figure 6.1*:

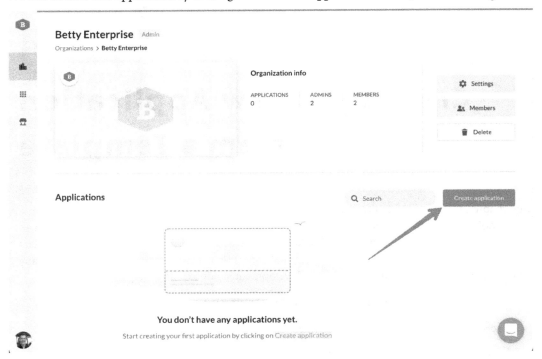

Figure 6.1 – Organization page

A page with an overview of all application templates will be visible, as seen in *Figure 6.2*. Here, you can find all public templates or any that have been created in your organization or others that you are a member of:

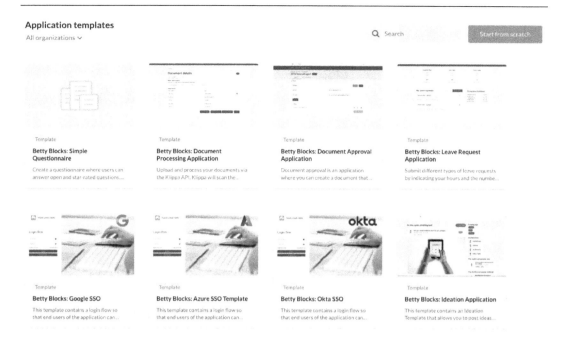

Figure 6.2 – Templates overview

The assortment of public application templates will keep on growing with the goal to give you enough templates to fulfill as many of your needs as possible. If you have a suggestion for a new template, you can tell the Betty Blocks team about it by clicking the **Request a template** button.

For now, we will create a *Questions and Answers* application. You can scroll down the list or use the **Search** field and search for Betty Blocks: Q&A Application. Selecting this template will give you more information about it with a description and an image. This is the case for all templates that are provided by the platform itself. The following screenshot shows the **Live preview** button, which you can use to see a preview of the template before deciding whether or not to use it:

Betty Blocks: Q&A Application

Template   By: Betty Blocks

This template contains a Q&A application where you can ask and answer questions. It also has a built in login flow. All the pages are built using the Page Builder. The template comes complete with models and data grids in the back-office.

Figure 6.3 – Q&A template details

This **Q&A Application** template gives you an application that is ready to go. It has pages and functionalities for users to register, reset their password, and log in. After logging in, they can browse already-asked questions or ask a new question. Clicking on a question will bring the user to a detailed page where all the answers are shown, and a user can add their own answer too. The user who asked the question can select one of the given answers as the best answer. This best answer will be added to the question on the home page of the application.

To continue, select **Use template** to base our new application on this template. A dialog will pop up asking for you to give your new application a name. It will allocate an identifier based on the name you enter. You can edit your own identifier if you like. Remember that the identifier has to be unique.

Click **Create application** to create the application (*Figure 6.4*). It might happen that the identifier has already been used. In that case, a message under the input will notify you about this. To solve this, you can either change the name of the application or fill in a custom identifier yourself.

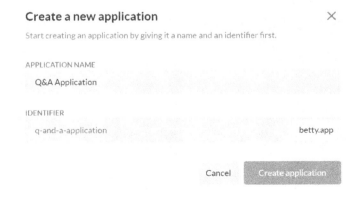

Figure 6.4 – Create a new application

When the name and identifier are verified with no errors, the application will be built for you. You will be redirected to your organization page. Here, you will see that the application is being created, as shown in *Figure 6.5*, and it will show the creation progress. When it is complete, only the title will be displayed in the white part of the application tile.

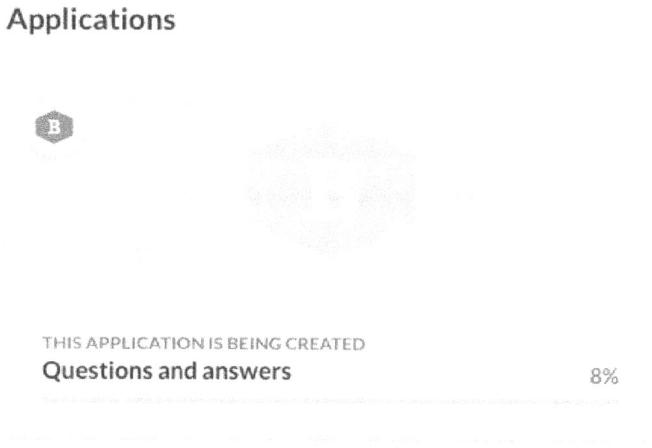

Figure 6.5 – Application is being created

We have successfully created our first application based on a template in the Betty Blocks platform. We will use this application in the upcoming chapters to walk you through the process of working on an application yourself.

Entering the development environment

Now that we have an application ready to go, let's take a look at the development environment of this application. The development environment of Betty Blocks can be reached by navigating to the application overview in your organization and then by clicking on the application tile with the **B** letter (for Betty Blocks) in it (see *Figure 6.6*) for the just-created application. It can be reached by navigating to the application overview in your organization and then hovering over the application tile (see *Figure 6.5*). Once you hover over it, you should be able to click on it. Here, you see more details about the application as walked through in Governance - Application:

Questions and answers

Organizations > Betty Enterprise > **Questions and answers**

Production application

STATUS
Enabled

● Development ∨

ORGANIZATION   LINK   ZONE
Betty Enterprise   Open application   nl3

👤 1   ⚒ 1   ...

Figure 6.6 – Application page

Since we want to focus on the development environment in this section, we will not bother creating sandboxes for the application for now.

There are two ways to open the development environment. The first and most obvious one is by clicking on **Open application** just under **LINK**. The second way is by hovering over the gray Betty Blocks logo and clicking on it. When you hover over it, it will become darker, and an **Open application** prompt will show. No matter which option you choose, the development environment will open in a new tab.

Now, we've arrived at the development environment of Betty Blocks. Let's see what we can find here.

Dashboard

Opening your development environment will always bring you to the dashboard of the application (see *Figure 6.7*). On this page, you can find helpful links to courses, tutorials, documentation, and the Betty Blocks forum:

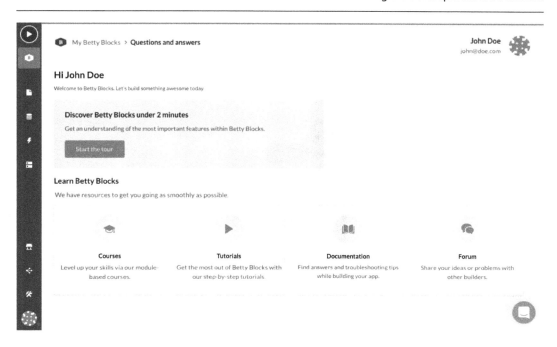

Figure 6.7 – Development environment dashboard page

All pages have the navigation bar on the left. We will walk through the top part of the navigation bar now. While developing the application, these parts of Betty Blocks will be used most frequently. The bottom part will be discussed at a later stage. The top part of the navigation bar consists of the following items:

- **Play** (compile)
- **Dashboard** (the Betty Blocks logo)
- **Pages**
- **Data model**
- **Actions**
- **Back office**

Here's a closer view of how it looks:

Figure 6.8 – Navigation bar

We'll be going through all of these parts in more detail now, so you'll get an idea of what they are used for.

Pages

Clicking the **Pages** option (see *Figure 6.8*) in the navigation bar will show a drawer with a list of the pages in your application, as seen in the following figure:

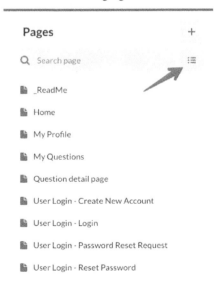

Figure 6.9 – Pages drawer

This drawer will give you most of the functionalities of the **Pages** section of your development environment.

Here, you can search for and/or select a page to open it in the page builder.

Clicking on the + button will show you an overview of different page templates. We will walk through this later in this chapter. As highlighted in *Figure 6.10*, the icon is for the list overview. We will click this one now to bring ourselves to the **Pages** page:

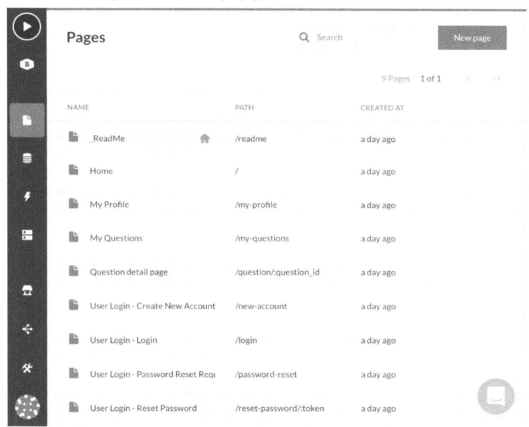

Figure 6.10 – Pages overview

The **Pages** overview (see *Figure 6.10*) lists the pages in your application with a bit more detail. It will also show the path of each page at runtime. Runtime means the version that your end users see in their browsers. Clicking this link will open the runtime page in a new tab. The last column shows you when the page was created. If someone else in your team is working on a page, their profile image will show up on the right of the **CREATED AT** column.

Next to the name of the page, there might be an icon. There are two icons that can be linked to any page of your application. The first icon is the *home* icon. This icon indicates the page that has been set as the home page. When the user does not have access to a certain page, they will be redirected to this home page. The second icon is the *404* icon. This icon indicates that the page has been set as the **Not found** page. When a URL cannot be found at runtime, the user will be redirected to this **Not found** page.

Hovering over a page in the table will make that row light blue and show extra actions on the right side of the row:

Figure 6.11 – Page options

The first one will open the page in runtime in a new tab. Clicking the second icon will duplicate the page. It will ask you to give this new page a name. The last icon will delete the page. After clicking it, a dialog popup will appear asking for your confirmation to delete the page.

At the top right of the page, you can find the **New page** button. Clicking this will show the templates overview, as shown in *Figure 6.12*. Here, you have the option to start a new page from scratch with a blank canvas or start with one of the page templates. These templates are designed to speed up the development of your application. Hovering over a template will show a bit more information about it, a button to start using the template, and a button to open a new tab with a preview of how the page would look after creating the page.

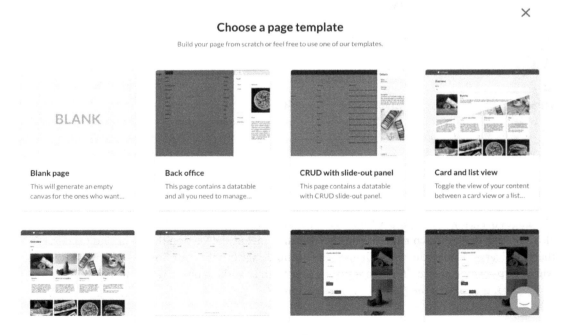

Figure 6.12 – Page templates overview

For now, we will close the page template overview by clicking on the **x** at the top right of your screen, as seen in the preceding figure. You will see the **Pages** overview again (*Figure 6.10*). Click the **User Login - Login** page to open it in the page builder.

Page builder

When you see the page builder, you should also see a login screen (see *Figure 6.13*). Let's look at this screen in more detail. Next to the navigation bar, you will see a new column with a lot of sections containing the page components in it. These components can be dragged onto your canvas, which is on the right of the screen where you can see the login page. We're not going to modify the page itself now, but we'll just go through it to better understand what we have here:

Figure 6.13 – Login page in the page builder

At the top of the components column, you can see four tabs, as shown in the following screenshot:

Figure 6.14 – Components, Variables, Component tree, and Settings

The first one, as we already saw, is the **Components** tab. This is the default tab that you see when you open the page builder. The second one is the **Variables** tab; this one requires a more detailed explanation with examples. We'll get back to this one in *Chapter 8*. The third one is the **Component tree** tab. This allows you to quickly see all the components with their hierarchy on your canvas and enables you to select the ones that might be hidden behind other components. Since a complete page might have 300-400 components, this can be very useful. And the last one is the **Settings** tab, which allows you to change the name of the page, the title, and the path for the page.

On the canvas of the page, let's click on the **Email address** input field. Once you select it, it should look like the following:

Figure 6.15 – The selected email address input

At the top of the purple box that surrounds the email address input, you can see an arrow pointing upward. Click on the arrow. You can also use your arrow keys (up and down) to navigate through the components. This redirects to its parent component. Basically, it's inside another component, and you have selected that component now. Now, you will have selected a **Box** component, as shown in the following screenshot:

Figure 6.16 – The selected box component

Click the arrow again, and you should have the form selected. Select the form. The form holds both fields that you see on your page, but it also does something specific: it's connected to an action. And with an action, you can do things such as write data to your data models or send an email, or maybe even call a web service. How can we see this action? Now that you have selected the form, you will see, on the left, that the components are gone, and it now shows a lot of options there. All of these options are the options for the form. Right now, we are only interested in the action that is behind this form. So let's click on the **Edit action** button:

Figure 6.17 – Edit action button

Now, you should see the action page of the platform, as seen in the following screenshot:

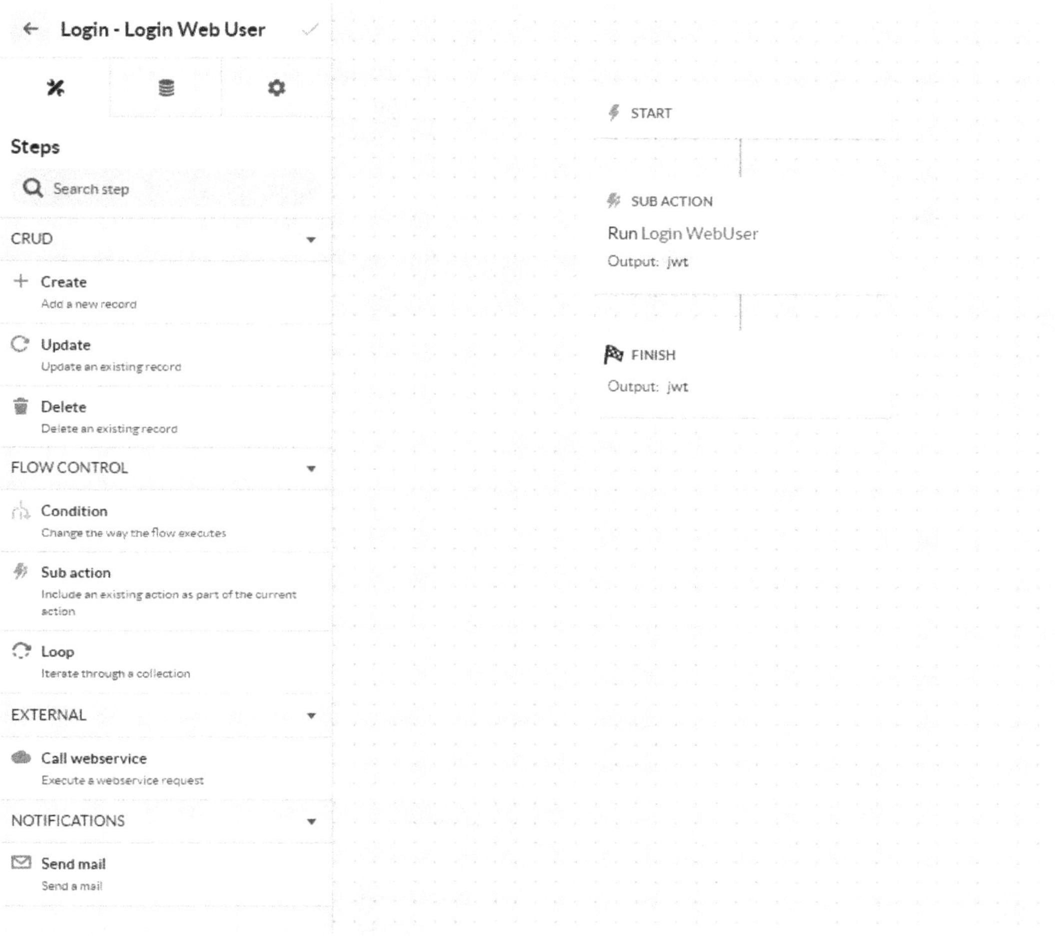

← Login - Login Web User ✓

✖   🗃   ⚙

Steps

🔍 Search step

CRUD ▼

+ **Create**
Add a new record

↻ **Update**
Update an existing record

🗑 **Delete**
Delete an existing record

FLOW CONTROL ▼

Condition
Change the way the flow executes

Sub action
Include an existing action as part of the current action

↻ **Loop**
Iterate through a collection

EXTERNAL ▼

☁ **Call webservice**
Execute a webservice request

NOTIFICATIONS ▼

✉ **Send mail**
Send a mail

⚡ START

⚡ SUB ACTION
Run Login WebUser
Output: jwt

🏁 FINISH
Output: jwt

Figure 6.18 – Betty Blocks action page

An action is a graphical and flow-based representation of a workflow that you would like to perform. In this case, the action is only doing one thing, and that is calling the login event (or step) to perform a login. All the logic is built into this; the only thing it needs is the email address and password and it will try to log the user in. If that doesn't work, it will return an error message to the page.

So, when you click on the **Login** button on the page, this action will be performed. As you can see, there is no code involved here. This part of Betty Blocks would normally be the backend code in your application, but it is all drag and drop instead. The menu on the left functions just like the page builder and allows users to drag and drop any event onto the action. It even allows you to make choices with the conditional event. But all of this will be explained in more detail in the next chapters. If you click on the arrow on the top left of the page, you'll be sent back to the page builder.

Let's click on the page builder icon in the navigation bar on the left side of the page. Then, from the drawer with pages, select the **Home** page. This should open the page shown in the following screenshot:

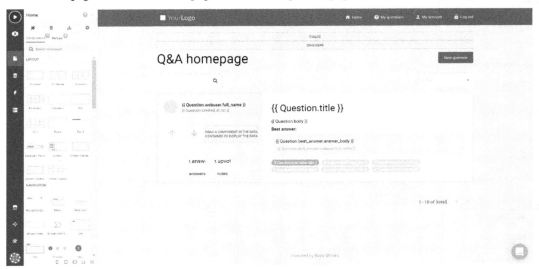

Figure 6.19 – The home page

Here, we have the option to create a new question by clicking on the **New question** button. This button opens a dialog box when you click on it. You can see the dialog as well. It is invisible on the actual page, but here in the page builder, it is still visible as a gray bar that allows you to select it very easily. Click on the gray bar that says **Dialog**; if you look on the left now, you'll see that the options appear for this dialog. There is a button there called **Toggle visibility**; click on it, and the dialog box should appear on the home page now. Now, in the dialog, you can select any component as well, just like on the rest of the page. Right now, you can't click on any of the components on the rest of the page because the dialog is blocking this. Just like with the previous form, let's click on the **Question** field and press the up arrow until we find the **CreateNewQuestionForm** component:

Figure 6.20 – New question form

Next, look at the options that open up again for this form on the left side of your page and click on the **Edit action** button again, as we did with the previous form.

You should see the action editor now:

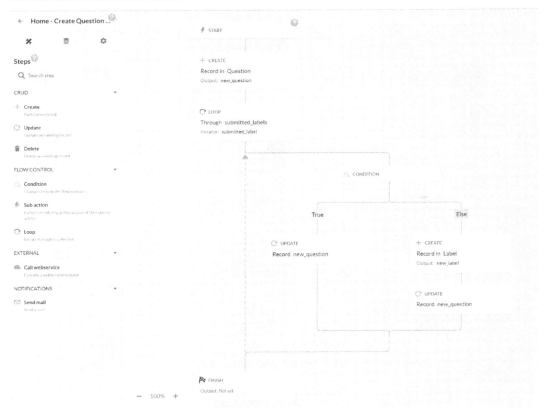

Figure 6.21 – Action for creating a new question

As you can see, this action is more complex than the login action that we saw before. Still, if you have never read code, I am pretty sure you can make out generally what is happening here just by looking at the visual way of how this is represented.

Let's go through it to understand it a bit better. The first thing we see in the flow is **START**. This is the point from which your workflow is starting. Then we see a **CREATE** event, which allows you to create new data in your data model. In this case, it will create a new question for us. Next, we see **LOOP**; this will go through a list of pieces of data and handle them one by one. This is what the next steps will do; they are inside the loop and will continue to be executed until the loop list is empty.

Next up is the **CONDITION** event, which can make a choice based on the data that you compare. In this case, it will check whether a label already exists in the data model. If it already exists, it will go into the **True** flow; otherwise, it will go to the **Else** flow. You can see that in the **True** flow, it will create a label, while in the **Else** flow, you only see the **UPDATE** step. This **UPDATE** step will update a question in the data model with the fields that you have filled in in the form. The last step is **FINISH**, which is where the action ends, and the users will be informed in the form whether the action was successful or not.

Let's click on the back button in the top-left corner to go back to the page. If you now click anywhere outside the dialog box, it should select **AddQuestionDialog**. This should make the **Toggle visibility** button appear in your options on the left:

Figure 6.22 – Button to toggle visibility

If you click this button again, it should turn off the dialog box.

We didn't go into too much detail because I'm just trying to show you how some of the basics of the platform work, to give you an understanding of what you can do. The last thing I want to show you in this chapter is the data model. We haven't seen where our data is actually being stored yet.

Data model

Essentially, your data model is in your database where all your data is stored. To view the data model, we'll need to click on the **Data model** icon in the navigation bar. This icon looks like this:

Figure 6.23 – Data model icon

As with the page builder, once you click on this, the drawer for the data model will open, as shown here:

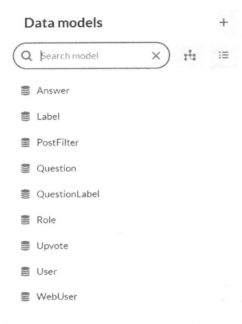

Figure 6.24 – Data models drawer

This allows you to quickly go to any model you want, but the data model has two different views as well – namely, the *schema view* and the *list view*. As the names suggest, you see a schematic view of the data model in the *schema view* and a list of all the models in the *list view*. Let's have a look at the *schema view* first. Let's click on the *schema view* button:

Figure 6.25 – Schema view button

As you can see in the following figure, you get a nice and clear schematic view of the data model:

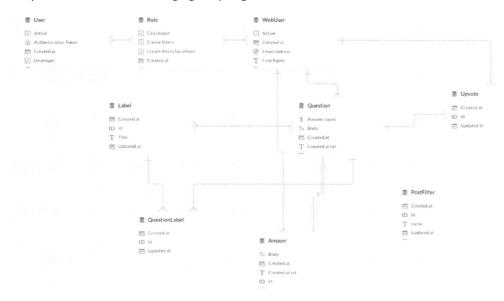

Figure 6.26 – The data model schema view

Each box in the schema is a model, which represents a data element inside the application, and has properties such as a name, description, date, or number. Between the models, you see connecting lines, which represent the relationships between the models. There are three different types of relationships on the platform. We'll go into detail about what types of properties and relationships are available in the next chapters. When you saw the **CREATE** or **UPDATE** event in an action earlier in the chapter, they referred to one of these models to either create a record or update a record here. We'll also dive deeper into this in the next chapters.

As you can see in *Figure 6.26*, the central part is the **Question** model in the data model, as it has the most relationships with other models around it. Remember in the **Create new question** action that a **CONDITION** event checked whether a label already exists? Here, you can see that it has a relationship with the **Question** model (the line between **Label** and **Question**). It might look a bit confusing at this point but it's actually just a translation of a real-life situation to a data model. In the next chapters, we'll be building our own data model step by step and explaining how to understand this. I hope, for now, it gives you a small impression of what the data model is. Let's have a quick look at the *list view* of the data model as well:

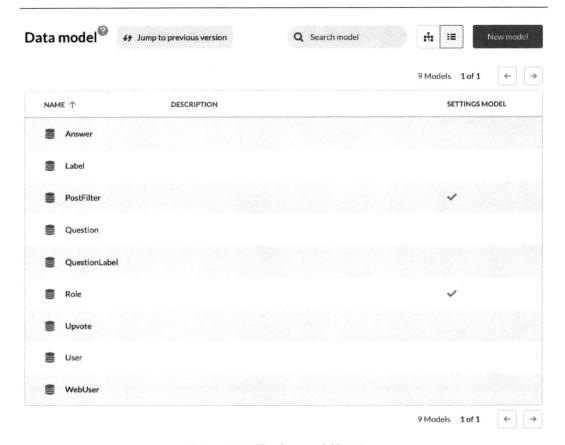

Figure 6.27 – The data model list view

The *list view* is exactly the same as the *schema view*, just represented in a list format in alphabetical order. If you have a lot of models, this might be a bit quicker to find what you need.

This was a quick and global introduction to the development environment of Betty Blocks showing some basic parts of the platform. Next, we'll go through the actual application that you've just installed using the Q&A template, before starting to build an application in the next chapter.

Launching the application

Now that we've toured the development environment of our newest application, it's time to see the actual application in action. The whole application will feel empty at this point; all the pages, actions, and data models are in place, but no user data is present yet. So while we explore this template, we will need to add data along the way, and that's also a great way to explore the template. Let's get started!

In order to get to the frontend part of our application, we'll need to launch it. There are several ways of doing this. The first and easiest way of doing this is by clicking on the *play* button (*Figure 6.28*) in the bar on the left:

Figure 6.28 – The play button as seen in the menu

This launches the application from the page that has been set as the home page. If you click on the **Pages** icon in the navigation bar and click on the **Pages** overview button (highlighted in the following figure), you'll go to the **Pages** overview:

Figure 6.29 – Pages overview button

Here, you can see an overview of all the pages, and one of them has a little house by it, indicating the home page:

Figure 6.30 – ReadMe with the home page icon

So, if we click on the *play* button in the bar now, it would redirect us to the /readme path. This is not what we want – we want to go to the **Home** page of our application. Right now, the easiest way to do this is by clicking on the **Open page** icon on the **Home** page row in the overview:

Figure 6.31 – The Open page icon

We'll show you how to change the home page in your application in the next chapter. Let's click on the **Open page** icon, and it should open the page for you in a new tab in your browser:

Figure 6.32 – The login page

As you might have noticed, you're not being redirected to the home page but instead to the **Login** page. Why is that, you might ask? Our home page is a secured page; you can't access it without being logged in to the frontend of our application. But we don't have a user right now, so we can't log in. Let's fix that first. Below the blue **Login** button, you see **Register new account**. Let's click on that, and it should take us to a new page where we can register a new account for our application. Fill in your details with your own email address and choose a password.

Now, you should be able to log in using the email address and password that you've just chosen. After logging in to your account, you should be able to see the home page:

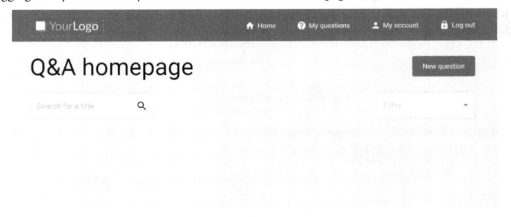

Figure 6.33 – The home page

As you can see, the home page is still empty, because the application has no data in it currently. So let's click on the **New question** button. It's the same button that we saw earlier when we visited the page builder – remember when we opened a dialog box? Let's click on this **New question** button, and we should get that same dialog box pop up. Fill in the form that is inside the dialog box. The last input is an auto-complete field for the label, but right now, there are no labels in the application; the labels are part of your application, so you need to add these to your database. You can add them by typing any name and pressing *Tab* when you are done. Click on **Submit**, and now the action that we saw earlier should run and add all this data to the data model. Your question should be visible on the page now:

Figure 6.34 – Your first question example

Click on the question title now (in my case, **My First Question**) to view the details of this question on the detail page. On the detail page, you can see that people can answer this question now by clicking on the **New answer** button. This opens another dialog box that allows you to add an answer to this question. Also, there is a small pencil that allows you to edit this question. Have a look at the development environment of Betty Blocks and see whether you can find the dialogs for these two buttons and view their actions.

Play around a little with this template to get a feeling for what it can do, and also try to see in the development environment how this was built. Hopefully, it will give you a bit more insight.

Summary

In this chapter, you've learned how to create an application from a template. We've used the Q&A template, so you can get a feeling of the platform by seeing an already-built application. We've toured all the major parts of the platform such as the page builder, actions, and data models. The tour gave an overview of these parts, and we'll go deeper into them in the next chapters. Finally, we've explored the application as a user, so we can see how it actually works.

Let's move on to the next chapter, where we'll make some changes to this application. We'll add some components to the page and change some data to help you to understand how all of this works.

Questions

1. How can you group your applications?
2. Which part of Betty Blocks is used for creating web pages?
3. What is the part called in which you can define your database in Betty Blocks?
4. What are sandboxes used for?

Answers

1. You can group your applications by using organizations. Organizations allow you to group your applications either by company, team, or in any other way you would like.
2. The page builder allows you to create pages using drag-and-drop components.
3. The data model. You can define the models and properties here that will allow you to store data.
4. Sandboxes are used to create production, test, and development environments, so you don't have to make changes in the environment that your users are using.

7
Prototyping an Application

As a citizen developer or no-code developer, you might need to convince your peers or manager about what you want to build. To do that, you can build a quick prototype. In this chapter, we'll go over how you can build a quick prototype in the page builder. This prototype will even save some data and show some data, so it will be much easier to convince everyone of what you want to build. We'll also connect these pages so that you can click on links or buttons that will take you to another page.

We will keep our use case simple for the prototype since you are at the start of your journey of building applications. In the few next chapters, we're going to keep adding more functionality to our applications. We'll build a home page where people can find some information, and we'll add a second page to that where you can fill in a form that will request information about what services the user wants to buy from the company. Finally, we'll add an overview page, where you can see an overview of all the requests that have been made through the form.

This chapter is divided into the following topics:

- Setting up the main page
- Partials
- Request page
- Request overview page

After completing this chapter, you should be able to set up a prototype with the platform and demonstrate a working prototype.

Setting up the main page

Let's create a new application in our organization in My Betty Blocks. To start with, it can be a blank application since we don't need any template for this use case. As always, you can choose your own name for your application. Once the application has been created, we can enter it and start setting up our main page.

Our main page is going to be a typical website page that contains information about the company. There will be the company's logo, a menu at the top, a header that will display some information about the company, and some additional information about the company. We can build this whole page from scratch or use the header and footer template. The header and footer template will give us a little head start, but since it is going to be a prototype to prove something, it doesn't have to be perfect yet; it just needs to prove a point. So, the quickest way to the end line is going to be the best here. For this use case, we can use the home page template. It will have most of what we need already in there; we just need to modify those things to suit our needs and maybe remove or add some extra stuff.

So, let's go to the page builder and click on **New page**. On the screen that opens, we can choose from multiple templates (*Figure 7.1*):

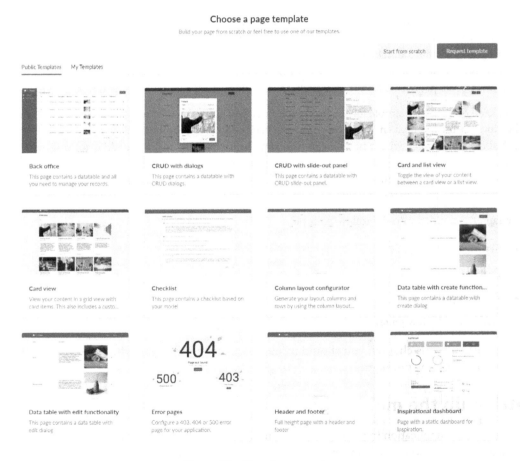

Figure 7.1 – Template overview

When we scroll down, we should see a template called **Page with homepage layout**. Select that one (*Figure 7.2*):

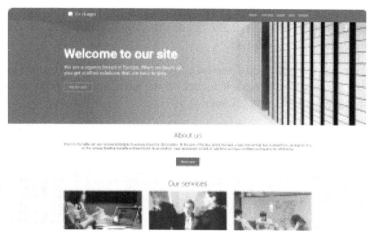

Figure 7.2 – Page with homepage layout

Let's name this page **Home** and use the same path for this page. We will also set the type of the page to **Public** since it will be our main page here, and we're not going to use authentication for our prototype.

Click on the **Create** button. Next, we'll set the configuration part of the template, which allows us to configure the partials for this page. We'll learn about partials in more depth later in this chapter, but in short, a partial is a part of your page that can be used on multiple pages and looks the same on all of those pages that you use this partial on; it also handles the same on all pages. For example, a menu is a partial as it will be the same on all pages.

Here, you can click on the **Auto generate partial** button for both **Top menu** and **Footer**. This will use a template for the partials as well. Now, let's click on the **Save** button to generate our page. Your page should look something like *Figure 7.3* now:

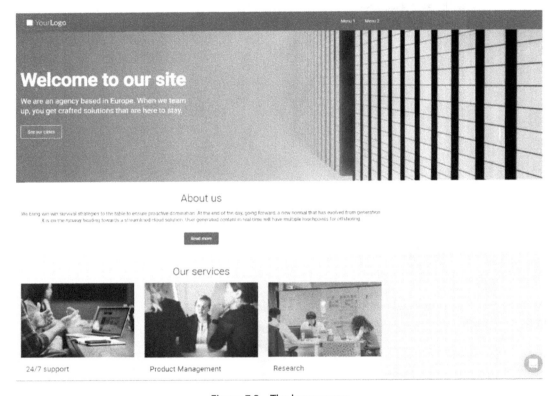

Figure 7.3 – The home page

If you scroll through the page in the page builder, you will see that this template has all the basics you would expect from a home page. It has a top menu, which you can configure, a header with a background image (also known as a hero area), and some short information about the page. There are also a few other sections below that, which you can customize to give the viewers of your page some more information about the company or product you want to promote on your page. Finally, at the bottom, there's the footer, which holds some links to other parts of the site and some information about the company.

For our prototype, we want to personalize the page a bit more so that it speaks to our audience. You can choose any topic that you would like to use. If you don't have one, then let's assume we want to create a page about solar panels where we will promote some solar panels. Then, we'll have another page where people can request an offer for solar panels.

Let's get started with personalization. First, click on the **Welcome to our site** text. This should open up the text component on the left, where you should see **CONTENT** in the component options (*Figure 7.4*):

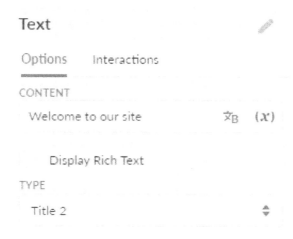

Figure 7.4 – The CONTENT option

Let's change this to **Welcome to Solar Inc**. As you will notice, the **Inc** part moves to the second line on your page. We would like to have all of this text aligned on the same line, so we need to fix this:

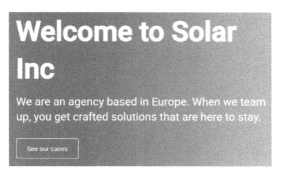

Figure 7.5 – Content on the next line

When you have the text component selected on the canvas, you'll see a little up arrow (*Figure 7.6*); click this up arrow to go to its parent component. With this arrow, you can always go to the parent component, which the component that you have currently selected lives in. This should select the column component:

Figure 7.6 – The text component with the up arrow icon

In the component options on the left, you'll see the column's widths, which are currently set to **6** for this column. That means it should take half of the space it has available within its parent component – as we learned in *Chapter 5*, *The Page Builder*, a column can have a maximum width of 12, so that makes 6 half (*Figure 7.7*):

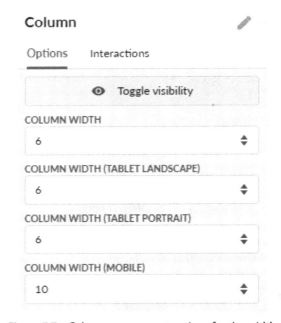

Figure 7.7 – Column component options for the width

As you can see, the options for the tablet have been set to **6** as well, while the mobile column width has been set to **10**, to make it work better for a mobile screen. In our case, we don't need a lot of extra space, so let's set all the **6** properties to 7. This should fix our text.

Let's click on the text under our welcome header text and change that to **We sell the most efficient solar panels in the country**. Now, let's change the image behind this text to make it appear more like a solar website. The easiest way to select this image is to click a little left or right of the text **Welcome to Solar Inc** until you select a box component that covers the whole image. Then, on the left-hand side in the component options, you should find an option called **MEDIA TYPE** under the **Background** category. Here, you will see a URL that refers to the image currently being shown:

Figure 7.8 – The background URL

So, let's get ourselves a nice image from the internet and use that for our page. You can use any image you want. I'll use something that refers to solar energy. Let's change the media type to **Image** if it's not set to that. This will allow us to upload an image to the platform that we can use on any page of our application. When you click on **Select image** (*Figure 7.9*), you'll see a button at the bottom of the opening field called **New file**. Let's click on that to upload the image since we don't have any images uploaded currently.

Select the image from your computer and save it to the platform. Afterward, you can select it in the **Select image** field by clicking **Select public file**. You should see the image change immediately on your canvas:

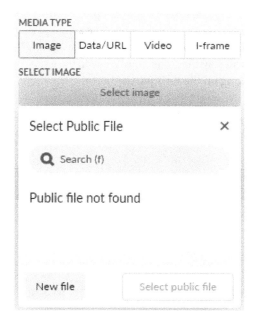

Figure 7.9 – The file upload select

So, now that we have made our first changes, let's do the same for the images in the three cards in the middle of the page. If click on the images on the page, you should see the images are set to their default. You can change the media type to an image if needed, or use the URL of an image to change it. Feel free to change some of the text on the page as you wish – just don't remove any parts of the page right now; we'll make some more changes to this page later, and we don't want anything to go missing.

Partials

The last thing we need to do on this page is edit the top menu bar. It's still standard, but if you click on it, you will see the borders are green instead of purple like the other components. This indicates that this component is a partial. So, what is a partial?

A partial is a component that can live on multiple pages but always looks the same on each page that you put it. So, when you change a partial, that change will take effect on each page that you use it on. That makes a partial ideal as a menu, for example, because it needs to look the same on each page that you use it on, or a footer, for example, because that is usually also the same on most pages.

If you want to edit a partial, you'll be brought to the partial editor. Since the partial itself can consist of different components, it needs to be edited in the partial editor. This ensures that the changes that are being made take effect on each page that it is used on. Let's make a change to our top menu partial.

After clicking on the top menu partial, in the component options, you'll see two buttons appear. First, you can convert a partial back into normal components. The actual partial will still exist, but for this page, it's being turned into normal components again, which also means that if you make a change to the partial, these components will stay the same now. We won't use that option here; instead, we are going to edit the partial by clicking on the **Edit partial** button:

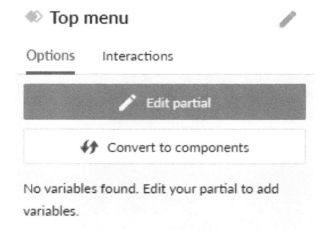

Figure 7.10 – Partial component options

With that, we've entered the partial editor; it's roughly the same as the page builder, except that it's zooming in on the partial itself and allows you to select all the individual components in the partial now. Let's click on the **Your Logo** image. Here, you can upload your own logo, which will appear on all the pages. If you have a logo, feel free to upload this here now.

On the right-hand side of the menu, you'll see **Menu 1** and **Menu 2**. If you select **Menu 1**, the components options will open for this button. Let's change **Menu 1** to **Home** and **Menu 2** to **Request**. If we click on the home button now, we can set it to point to the home page. You can change this by setting the page it should redirect to. To do so, click on **Select a page** and choose the **Home** page:

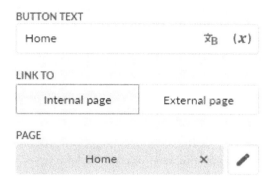

Figure 7.11 – The button's component options

This button will now redirect you to the home page every time you click on it, as well as on all pages where you will use this partial. This means you will never have to worry about having to set it up again. There is still the second button, but since we don't have a second page yet, we'll come back to that partial later and edit it. To exit the partial editor, you can click on the little back arrow in the top left corner of the page builder. It should take you back to the home page in the page builder. You should also notice that all the changes that we've made in the partial editor are shown now on the home page.

Now, let's create the request page.

Request page

The request page will have a form in it that will submit the fields from this form to the data model. This way, you can show some interactivity from your page as well. On the next page, we'll show all the results from that form in an overview.

Let's create a new page. Go to the page icon in the builder menu, click on it, and select the new page icon in the opening drawer. You should see the page template overview again. For this page, we'll use the header and footer page template since we want to reuse all of the partials that we modified in the previous section:

Header and footer

Full height page with a header and footer

Figure 7.12 – The header and footer template

The header and footer template is the same as an empty page, except it has the header and footer predefined on it. It makes building the page a little faster. Select the header and footer page template and name it **Request**. Set it to public again, just like the last one. After this, you will see that the partials have been automatically selected for you. These are the ones that we used on the previous page as well, so we don't have to do anything to apply them to this page. So, let's create the page now.

In the middle of the page, you'll see a box. Let's drag a column inside this box so that we can center our form more easily later. Select the column. In the purple box that surrounds the column, you'll see an arrow pointing up, similar to the one we clicked on previously to get to the parent component. In this case, we want to select the row component.

So, when we click on the up arrow, we should have the row component selected. This allows us to set the width of the row since just adjusting the column will make it smaller. However, this is not the case for the middle of the page. Here, we need to change the row. The row has component options on the left for doing this easily (*Figure 7.13*). The width here is set to **XL** by default, but for our purpose, **S** will do nicely. So, let's set it to **S**; it should become a lot smaller. This will allow us to put a form inside here and make it appear not too wide for our users:

Figure 7.13 – Width of the row component

Let's start with our form. It will need a data model to save the data. You can create a manual form without it being attached to any data model so that your data won't be saved. This would work fine for our prototype, but since it's almost just as easy to create one connected to a model, let's do that. First, let's think about what we want the form to do. Since we are building an application for a solar energy company, we want it to reflect something that makes sense in that use case. Users should be able to enter their first and last names and their email addresses, request what kind of solar system they want, what type of roof they have, and how many solar panels they would like to have. So, we will need the following fields:

- **First name** (text)
- **Last name** (text)
- **Email address** (email)
- **Solar system** (list)
- **Type of roof** (list)
- **Number of solar panels** (number)

Now, we can do one of two things: we can go to the data model and create our model and properties first or we can create it in the page builder. We will use the page builder here so that you know what to do.

Go to your component overview on the left-hand side of the page builder and type form in the search field (*Figure 7.14*). Then, drag the **Create Form** component inside the column that we just dragged onto the page.

Figure 7.14 – Searching for the create form component

Click on the **Select model** button. At the bottom of the box that appears, you should see a button called **new model**; click on it. Name your model `Request` and add the properties according to the list provided previously. A dialog box will appear, as shown in Figure 7.15:

Configure form fields   ✕

SELECT MODEL

Select model

SELECT PROPERTIES

Models	NAME	RELATION	MODEL	
No models found				

Cancel   Save

Figure 7.15 – The Configure form fields dialog

After adding the model and properties, your select model box should look like what's shown in *Figure 7.16*:

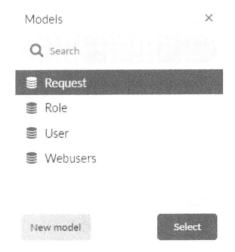

Figure 7.16 – Selecting the model box containing the Request model

Now, we can select the **Request** model. Click on the arrow next to **Request** in the **SELECT PROPERTIES** box. This should make all the properties appear (*Figure 7.17*):

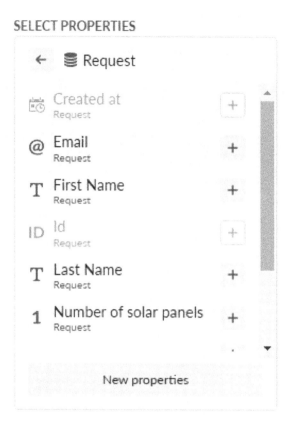

Figure 7.17 – The SELECT PROPERTIES box

Now, click on the + signs next to all the properties so that we can add them all to the form. Try to select them in the right order; this will put them in the correct order straight away. At the time of writing, it's not possible to drag them in the correct order in this dialog. This might be the case by the time this book gets released. Once you are done, hit the **Save** button; a form will be created for you. Now, your page should look like what's shown in *Figure 7.18*:

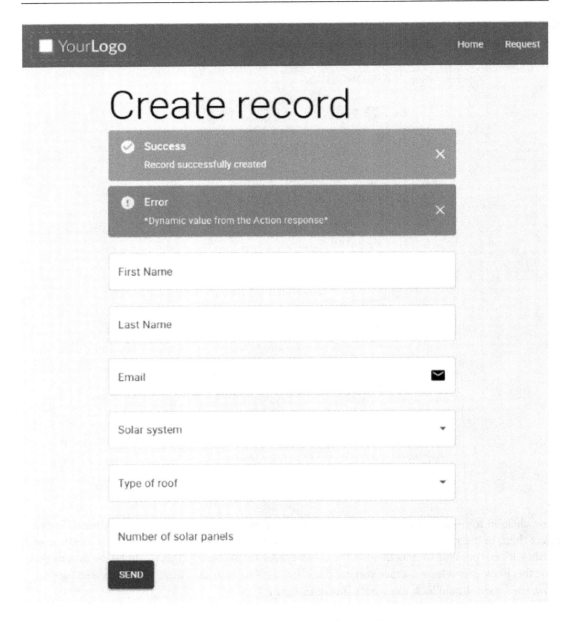

Figure 7.18 – The page with the form added

Click on the **Create record** and change the text to Request quote. Also, let's change the button at the bottom. It says **SEND**; we would like it to say **Submit request**. Select the button and, in the component options, change the button's text to Submit request. If you like, you can also click on the icon to add a nice little icon to your button.

Now, we only need to do one more thing: get this working properly. Since we've added the list properties through the page builder, the list properties don't have any values. This is because we couldn't add those in the process of creating them. So, we'll need to go to the data model to do that. Click on the data model icon in the builder menu and select the request model there. Then, click on the **Solar system** property so that we can modify it. A sliding pane should open on the right-hand side of the screen, with a button that says **Add value**. Click on it. Here, we can add values to the solar system. Let's add the following three values:

- **Solar 2000**
- **Sun Energy**
- **Betty Solar**

Every time you want to add a new value, click on the button again. Once you are done, at the bottom right, click on the **Save** button. Let's do the same for the types of roofs. Add the following two values there:

- **Gable Roof**
- **Flat Roof**

Don't forget to save it again. Then, we should be done. Let's go back to our home page first so that we can edit our partial. We can also do this from our request page, but we haven't published it yet, which means we can't access it. So, we need to do that too.

When you open the home page, open the top menu partial again by clicking on it and then clicking on the **edit partial** button. Click on the **request** button in the top menu. Select the page it should go to when you click on it by setting the page component option to the request page (*Figure 7.19*):

Figure 7.19 – The page set to the request page

Exit the partial editor by clicking on the back arrow in the top left corner so that we end up back on the home page. Now, we can click on the play button in the builder menu to publish our page. After a few seconds, a second tab should appear in your browser; you should see the page you've just created. So, every time you make a change to a page, you need to publish it first before you can see those changes on the actual page. With that, your home page is live on the web. You will notice that if you click on the request button in the top menu now, you'll get an error message because we haven't published that page yet. So, let's go back to your Betty Blocks tab and open the request page in the page builder. Hit the play button again once it's open to publish the request page as well. Now, you can click on both buttons in the menu to go to each page.

Now that you can also test out your form, let's enter some random data in the form a few times and submit it. You'll notice it will tell you that it has submitted the data from the form every time you click on submit, but how can we access it? We'll create a page for this in the next section.

Request overview page

The last page we'll create for this prototype will be the overview page for the requests. I would like to start with a little warning here – the way we are going to create this page is not the way you should normally be exposing your data. By this, I mean that we'll be publicly displaying data on a page without any form of authentication. This could be the case in certain use cases, but always be very careful when exposing your data to a public and open page. This is because you will be exposing a complete model to that page, not just the properties you will be showing on the page, so anyone smart enough can query the rest of the properties as well. The other models in your data model should be fine – they all have their own permissions, so you can only expose one specific model. But I just want to make you aware of this and that Betty Blocks has a lot of safety features to make sure you don't do this by accident, as you will notice in this section.

Like in the previous sections, we'll start with a new page again, this time based on the header and footer template again. There are templates out there that can do most of this as well, but for you to experience this for yourself, we'll do it with components instead of a template. In this case, this isn't that much more work.

So, let's call our page **Request overview**, set it to public, and select the default partials again, like last time. Then, drag a column onto the page again. Then, in the components area, search for the **DataTable** component. Drag the **DataTable** component into the column; a dialog should appear. You'll need to select the model again like last time, so let's select the **Request** model and then select all the properties that we want to show. Select all the properties that we've created. Remember that the order in which you choose them is the initial order that they will appear. You can always change the order later (*Figure 7.20*):

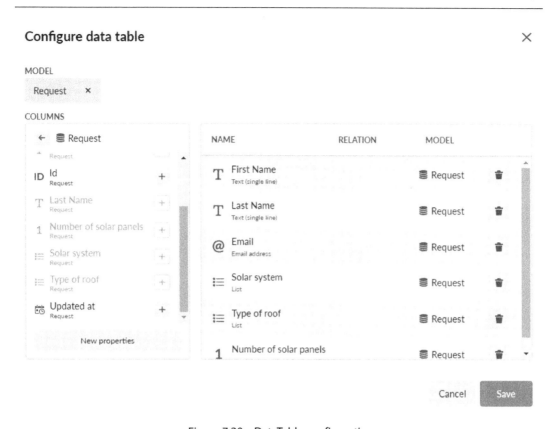

Figure 7.20 – DataTable configuration

Hit the **Save** button; your **DataTable** component should be created for you. Your page should now look like what's shown in *Figure 7.21*:

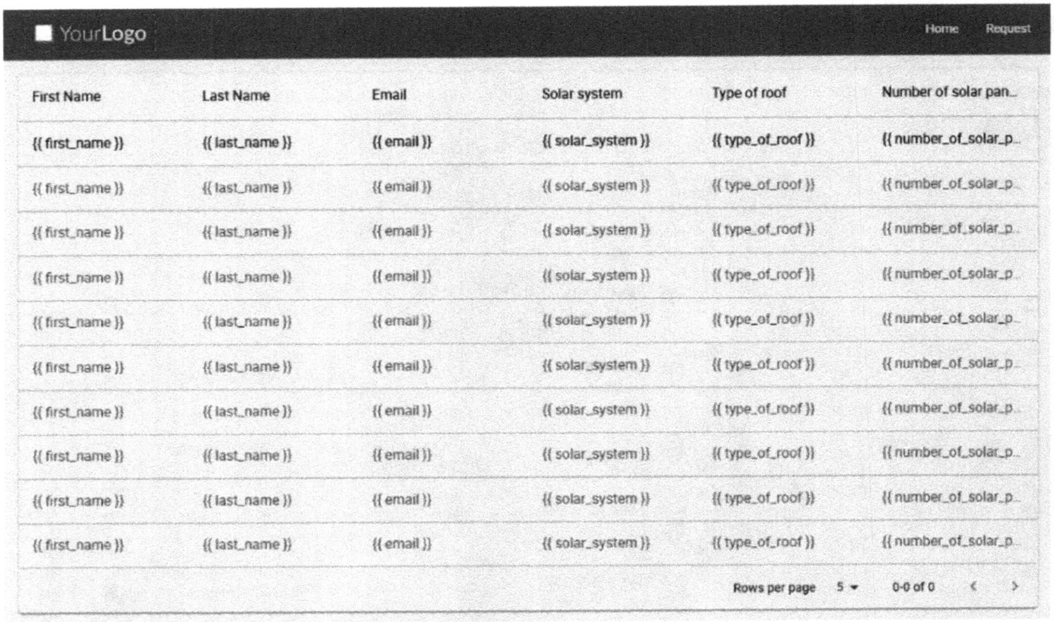

Figure 7.21 – Your overview page with the DataTable component

That is how easy it is to create a page that provides an overview of all the records in your model. Let's hit the play button in the builder menu to publish our page. As you may notice, you'll get a Betty Blocks login screen again. This has to do with the fact that you are trying to display data from your data model. An extra layer of protection is turned on by default so that no data can be accessed without there being a Betty Blocks user in your application. So, if you log in with your Betty Blocks credentials, you should be redirected to the overview page.

Now, you'll notice a strange message in the **DataTable** component – it's not showing any data. This is because our model doesn't have permission to show this data publicly. We can change this for this prototype. In Betty Blocks, click on the data model icon in the builder bar and open the request model. Click on the **Permissions** tab next, and then click on the **Go to roles and permissions** button. Note that this button might not be present at a later stage, so if you don't see it and you see an overview of all the permissions, then you are in the right spot already. You should see what's shown in *Figure 7.22*:

Roles and Permissions

Permissions determine what a user (based on the given role) is or isn't authorized to do. Learn how permissions work

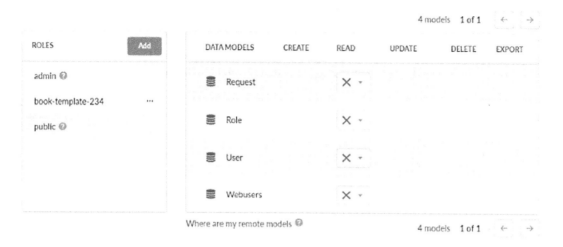

Figure 7.22 – The permissions overview

On the left-hand side, in the **ROLES** box, select **public** and then set the read permissions for the request model to allowed so that it's allowed to show its data publicly. Once again, be careful with this, and don't do it just for any model. Now, go back to the tab where your request overview page is and refresh that page. You should see all the data that you entered earlier:

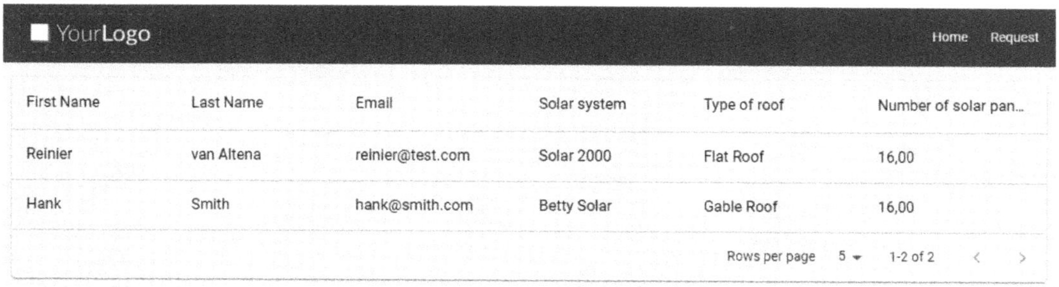

Figure 7.23 – The DataTable component with our data

That's it – you've created your prototype. If you want, you can add an extra button to your partial so that you can access the overview page more easily. Play around with that a little to get more familiar with Betty Blocks. In the next chapter, we'll build a to-do application from scratch so that you become more familiar with pages, actions, and data. We'll also cover creating, editing, and deleting data.

Summary

In this chapter, we built a prototype application and built a home page based on a template of our home page template. We modified this home page to learn about modifying text and uploading images. After that, we looked into partials, which allow you to quickly add menus to several pages without you having to modify them on each page individually. Then, we added our first form to the next page, which allows you to add data to your data model. Finally, we created a page with a **DataTable** component, which allows you to view the data that you've submitted through the form on the previous page.

In the next chapter, we'll create a to-do application. This will teach you more about creating, updating, and deleting items from the data model in the page builder. It will also cover interactions for the first time. These interactions are very useful for making your page more interactive.

Part 3: Building Your First Application

In this part, we'll start building our first application. This will be a simple To Do list application.

This part has the following chapters:

8

The To-Do Application

In this chapter, we'll focus on building a simple to-do application. This should familiarize you with all the basics that you've been taught in the previous chapters. It will be an application that will make use of a data model, the page builder, and actions for our logic. We won't use any kind of authentication in this application. We'll focus on that in our next project.

This chapter is divided into the following parts:

- Setting up the data model
- Building our to-do overview page
- Adding a to-do item to our list
- Our first interaction

After this chapter, you should be able to build a simple application yourself, where you can store data and configure interactions with your components. In the next chapter, we'll add functionality for updating and deleting data in this application. While building this application, you'll also get more familiar with some other options in the page builder and the use of components such as the data table, and we'll also focus a little on making our to-do list look presentable by configuring some styling options for our components.

Setting up the data model

As you've seen, there are multiple ways of setting up your data model: you can do it from the page builder as well as while building your page. I'd recommend starting with designing your data model as much as you can before building the pages. This doesn't mean you have to build all the models and properties at the start of your project, but having a good outline can help you quickly catch any mistakes, so you don't have to repair these later. You can always add more properties or models as you move along with your project, of course – it doesn't have to be perfect from the start.

In this project, we'll build a to-do application. This application will store a list of tasks that users have to perform.

Let's get started by creating our task model. For now, we'll create it in the data model view, but it's also possible to create your data model while creating your pages. We'll come back to this second option later.

Opening the data model builder

In the builder menu on the left side of your screen, as shown in the following figure, you'll see a white little database icon. When you hover over it, it should be highlighted in purple. Click on this database icon, and a quick overview of the data models should open.

Figure 8.1 – The data model option in the builder bar

After clicking on this button, the following page will appear:

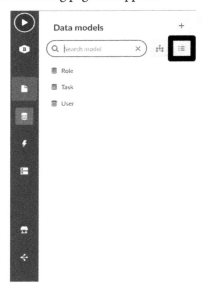

Figure 8.2 – A quick overview of the current data models in your application

This quick overview is great for quickly opening a model you are looking for or equally quickly creating a new model. But in this case, we want to go to the model overview page. We need to click on the overview button. This is the gray button with a list icon on the right of the search bar (see *Figure 8.2*).

Click on the list icon to open the **Data model** overview page, which looks like this:

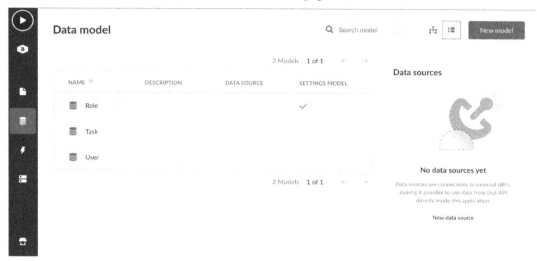

Figure 8.3 – The overview of the currently existing models in the application

On the **Data model** overview page, you get an overview of all the models you've created so far. In our case, this model contains only the necessary Role and User models, since we haven't created any other models yet.

Figure 8.4 – Button for showing all the models in a list view

There is also an option to search for your model by name. This should make it easier to find your model when they start to accumulate.

Creating the Task model

To create a new model, simply click the **New model** button. You can give your model any name you like. It's best practice to provide a name in English and singular form. We'll be using the name Task for our model. Once you've entered the name, simply click on the **Create model** button to create the model (see *Figure 8.5*).

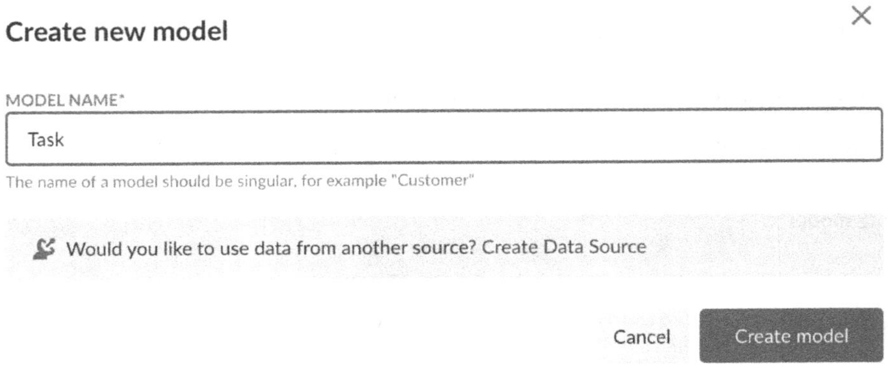

Create new model

×

MODEL NAME*

Task

The name of a model should be singular, for example "Customer"

> 🖋 Would you like to use data from another source? Create Data Source

Cancel   Create model

Figure 8.5 – New model creation

The new model will be created and will open on the screen, showing the basic properties created by default. Now let's add our own properties. We can do this by simply clicking on the **New property** button.

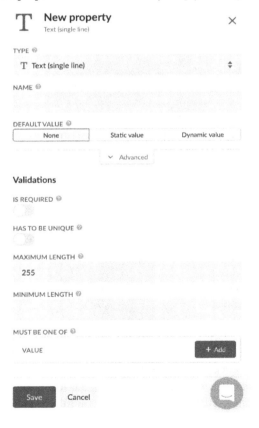

Figure 8.6 – Screen for creating a new property

In the properties options in the preceding figure, we can define the name of our property and select a specific type for it. We can also configure whether it should be required or unique (see *Figure 8.6*).

Let's create the following properties:

Name	Type	Required
Name	Text (single line)	Yes
Description	Text (multiline)	No
Status	List	No

The Status property will contain the following values, just as shown in *Figure 8.7*:

- To do
- In progress
- Done

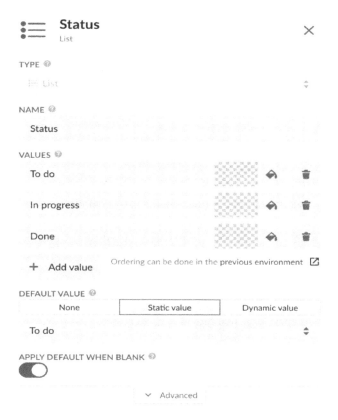

Figure 8.7 – The Status property with the selected settings

Make sure that under **DEFAULT VALUE**, you select **Static value** and choose **To do** from the dropdown for the default value (see *Figure 8.7*). In here, we can select the status.

In order to show our data online, we also have to set permissions for our new model. Click on the **Task** model and select **Permissions** to go to the permissions overview (see *Figure 8.8*).

Figure 8.8 – Open the Permissions overview

You will see the following overview of roles on the left of the screen. Let's select the **public** role and set it to **Allowed** for our **Task** model (see *Figure 8.9*).

Figure 8.9 – The permissions per role per model

Now we've finished creating our model and properties and setting the permissions, we can start creating our pages.

Creating the Task page

In this section, we will create a task page that provides an overview of tasks. We'll add the functionality to create a new task as well.

Opening the page builder

In the builder menu on the left side of your screen, you'll see a little white page icon. When you hover over it, it should turn blue, just like in the following figure. Click on this icon, and the **Pages** quick overview should open (see *Figure 8.10* and *Figure 8.11*).

Figure 8.10 – The builder bar with the pages icon selected

This quick overview is great for quickly opening the page you are looking for or quickly creating a new page.

Figure 8.11 – An overview of your current pages

But in this case, we want to go to the main **Pages** overview. Therefore, we need to click on the overview button. This is the gray button with a list view icon next to the search bar, highlighted in the preceding figure. This will open the **Pages** overview shown in the following figure:

Figure 8.12 – The overview of your current pages

On the **Pages** overview (*Figure 8.12*), you can see all the pages you've created so far. In our case, the list is still empty, since we haven't created any pages yet.

There are three categories of pages, as you can see along the top of the window:

- **All**
- **Public pages**
- **Authenticated pages**

Public pages can be viewed by anyone without being logged in, while authenticated pages require users to be logged in first before they can access the page. And of course, under **All**, you can find both of these pages.

There is also an option to search for pages by name. This should make it easier to find your page once they begin to accumulate.

The **New page** button allows you to create a new page. So let's start there.

After clicking on the **New page** button, we'll have to select a **Page** template. In our case, we'll select the **Header and footer** template (see *Figure 8.13*), and click the **Use** button.

Figure 8.13 – Header and footer template

Now we'll have to give the page a name. Let's call it Tasks and set the type of page to **Public**. After clicking **Create page**, we'll have to configure our header and footer. We'll skip this step for now and click **Add without configuration**. With that, we've created our first page.

Creating an overview of tasks

Now that we have our page, we want to create an overview of our tasks and show it in a table with the current description and status. To do this, search for the datatable component and drag it onto the page (*Figure 8.14*).

Figure 8.14 – Searching for the datatable component

We will drag this item into our Box component, which is already present on our page (see *Figure 8.15*).

Figure 8.15 – The Box component on your page

After dragging the component onto the page, we'll be presented with a wizard to configure our data table and select the model for which we want to show the data (see *Figure 8.16*). Let's click the **Select model** button and select our **Task** model. After this, let's click on the **Task** model and select the properties we want to show on our page.

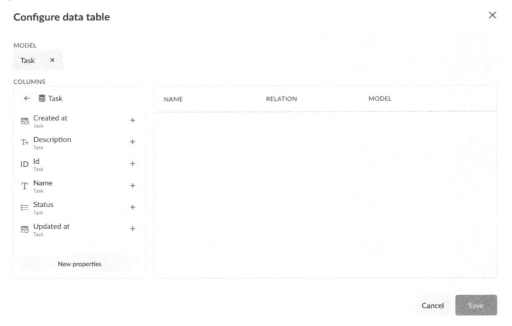

Figure 8.16 – The initial property selector for our data table

Then the property selector shown in the following figure should be filled with all the properties we want. Let's add **Name**, **Description**, and **Status** and click **Save** (see *Figure 8.17*).

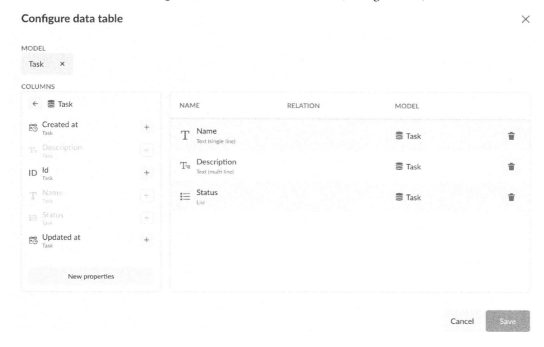

Figure 8.17 – The property selector for our data table with the selected properties

Our data table is now created and will look like this:

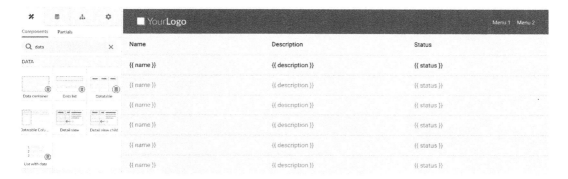

Figure 8.18 – The generated data table linking to the Task model

If you now compile the page by clicking on the **Play** button, you will see that currently, no data is present in your data table. So let's create a form that will allow us to create new tasks.

Creating a new task

In our component set, we'll search for the **Create Form** component (*Figure 8.19*) and drag it onto the page.

Figure 8.19 – The Create Form component

After dragging it onto the page just above our data list, a wizard will be shown. In this wizard, we can configure our form and select the model for which we want to create new data. Let's click the **Select model** button and select our **Task** model.

Now, let's click on the **Task** model and select the properties we want to show on our page. Let's add the **Name**, **Description**, and **Status** properties, then click **Save** (see *Figure 8.20*). Our new form is created, including the required action.

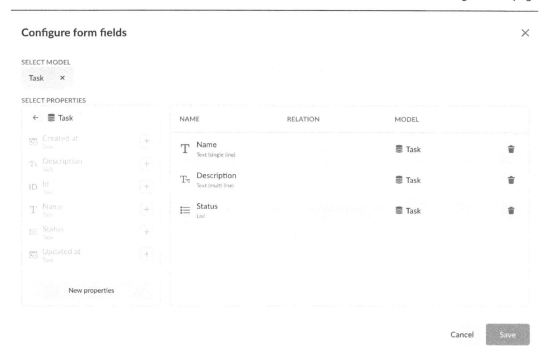

Figure 8.20 – The property selector with the selected fields

Let's compile the page button by clicking on the Play button. You can now create your first task in our created form. You have probably already noticed that your created task doesn't show up in the table, but it will after refreshing your page. This is where interactions come in to allow us to fetch the data in our data table after creating a new record. In the next section, we'll discuss interactions and add them to our page to make it more interactive.

Adding our first interaction

Component interactions enable you to dynamically change the contents of your page by adding more flow to your application. For example, interactions are helpful when we want to automatically refetch data after submitting a form. Refetching means loading data from our data model again after a change has been made, for example, to a page.

The Betty Blocks platform provides predefined options for setting up interactions. These are also used, for example, to show and hide specific components. For this we have the show-and-hide interaction. It can hide a whole column with all of its components. This interaction is placed on a button, so that when you press the button, the column is hidden or shown, depending on your needs. There is also a Hide/Show interaction, which will allow you to show or hide something when you press a button, for example. This means you don't have to create two buttons, one for hiding and one for showing.

For this use case, we'll use the refetch interaction. This will refetch data from our data model after the form we've created has been submitted – the action. Once that action has been completed, it should refetch the data from our data list, so the user will almost instantly see the result in their data table.

Let's select our **Create form** component in the component tree. Click on the three dots and select **Options**.

Figure 8.21 – Component tree with Options shown

Inside the configuration options for the **Create form** component, click **Interactions**, after which you will see the following screen (*Figure 8.22*):

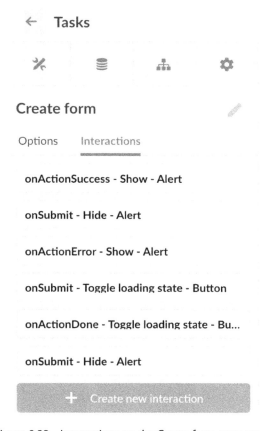

Figure 8.22 – Interactions on the Create form component

There are six different interactions here, which have been created for you by the Create form component that you added earlier. So, what does all this mean? Let's take the **onSubmit - Hide - Alert** interaction. The first word, **OnSubmit,** means that the interaction will fire an event when the form has been submitted. This will *hide* something, which is the second word we see here. What will it hide? It will hide the **Alert** component, the third word we see here. This approach quickly explains what each interaction in this list will do. Let's create our own interaction now, since we need the form to refetch the data from our data table.

Select + **Create new interaction** and then select the **More** option. Here, we have to specify that when the action has been successfully finished – **onActionSuccess** – we want to refetch our data table.

We can do this with the following three simple steps:

1. Determine the event (**Click**, **Change**, or something else).

2. Select the function for your component. The specified function will be performed after the chosen event has occurred. This could be a refetch, hide, or show, for example.

3. Select the component you want to target with your function.

This will give you the following configuration:

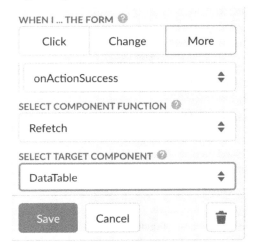

Figure 8.23 – The interaction for refetching data from the data table

Let's click **Save** and recompile the page. Now we've created our first page, and after submitting a new task, it will show up in our data table.

Summary

In this chapter, we made a start on our to-do application. We added our data table, which fetches data from our data model. We created a form that allows us to add data to our data table. We also encountered interactions for the first time, which allow you to dynamically change components on your page by configuring an event that will fire on a specific occurrence, for example, clicking a button or the completion of an action. Then the interaction will perform a function targeting a specific component. The function could be showing, hiding, or refetching data from the data model.

In the next chapter, we'll learn how to add a *create, update, and delete* flow to our application. We've now set up the basics for this application.

Questions

1. What is an interaction used for?

2. Which triggers for interactions did we cover in this chapter?

3. Which components can an interaction target?

4. Why do we need to set our permissions on the data table?

Answers

1. Interactions are used to add dynamic changes to your page without having to reload the page.

2. **Click**, **Change**, and **OnActionSuccess** triggers.

    Click triggers activate when you click on the given component. Change triggers work when someone types in a field, for example. Of course, there are a lot more interactions available beyond just these three.

3. Interactions can target almost any component. Usually, the target type has been defined already by default, so you don't accidentally target something that would be invalid.

4. We need to set permissions so data is retrieved securely. If a specific role isn't allowed to view some data, then that data will not be retrieved and an error will be shown in the component.

9

The ToDo Application – Actions and Interactions

In this chapter, we'll focus on extending the functionality of the to-do application. This should reinforce the basics that you learned about in the previous chapters. The application will make use of a data model, the page builder, actions, and interactions for our logic. We won't use any kind of authentication in this application. We'll focus on that in our next project.

This chapter is divided into the following sections:

- Viewing data in a dialog

- Editing data in a dialog

- Deleting a record

By the end of this chapter, you will have a better understanding of how to view, edit, and delete the data in your application. While building this application, you'll also become more familiar with some other options in the page builder, by using components such as Data Tables and Dialogs. We'll also focus a little on making our to-do list look presentable by configuring some of the styling options of our components.

View tasks

In the previous chapter, we created a basic setup for our application. We will extend this in the current application. The first thing we will do is view the details of a task.

Let's get started by creating our first dialog. First, we need to open the **Tasks** page we created before.

Your page should look like this:

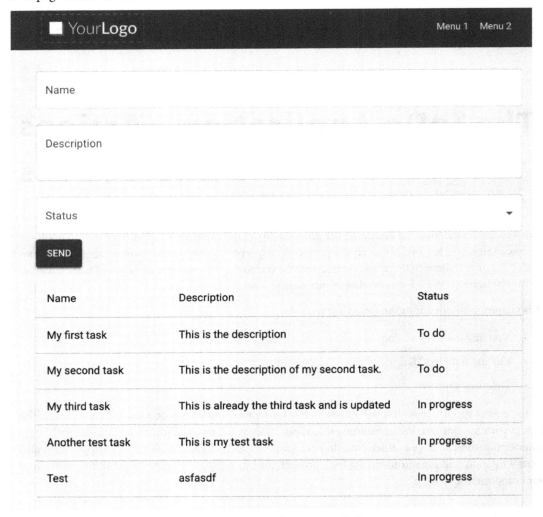

Figure 9.1 – An overview of our current Tasks page

Let's search for a column in our components and drag it just above the **Create form**:

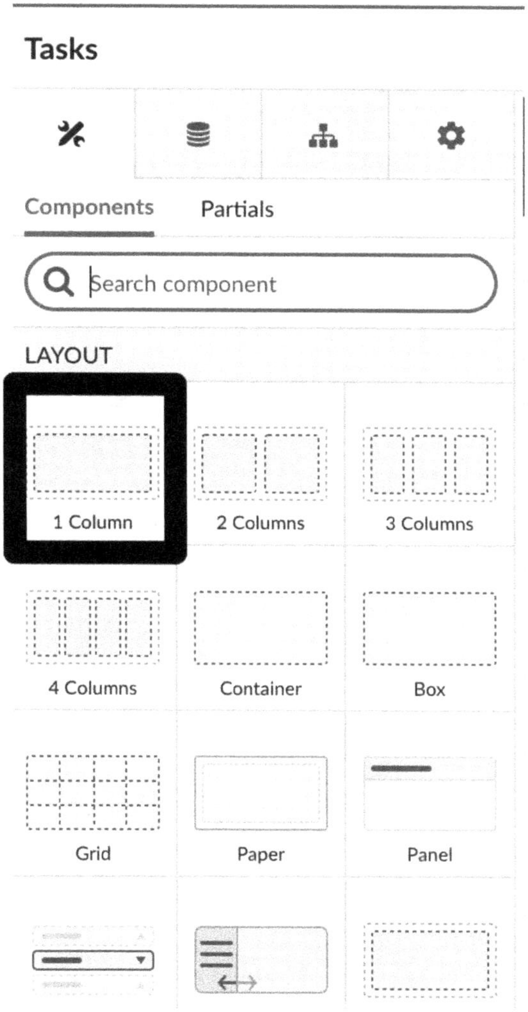

Figure 9.2 – A column component

After dragging this on the page, you will end up with the following:

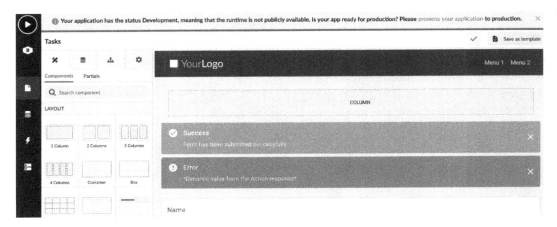

Figure 9.3 – The Tasks page with a new column above the Create form

We will now search for a `dialog` component and drag it onto our new column:

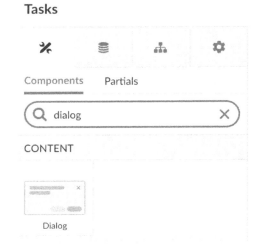

Figure 9.4 – The Dialog component

Now, let's select the **Dialog** component and give it a new name, so we can easily recognize it when creating our interactions:

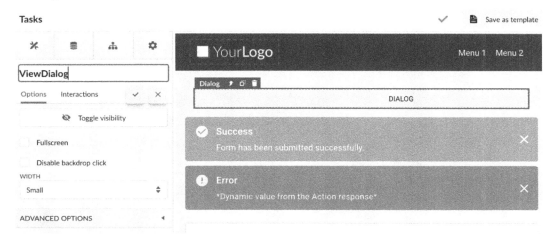

Figure 9.5 – Renaming our new dialog

Now, we want to change the look and feel of the dialog we've just created. In order to be able to do this, select the dialog and turn on **Toggle visibility**:

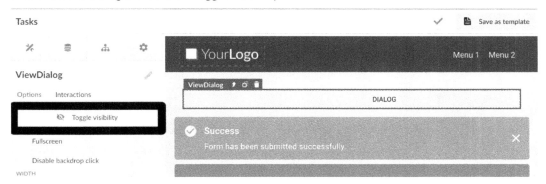

Figure 9.6 – The button to toggle the visibility of a dialog

Our dialog will pop up, and we can change the content:

Figure 9.7 – A basic dialog

For the first basic setup of our dialog, we will delete the **SUBMIT** button, change the name **Dialog** to **Task details**, remove the text block, and change the text of the **CANCEL** button to **CLOSE**:

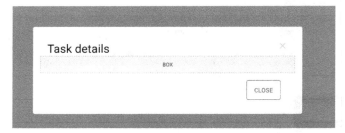

Figure 9.8 – The dialog form after the changes

Now, we'll search for the `detail view` component, and we will drag it into the box:

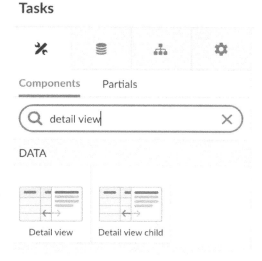

Figure 9.9 – The Detail view component

After dragging it in the box, a wizard will appear, in which we can select the model for which we want to show the detailed data:

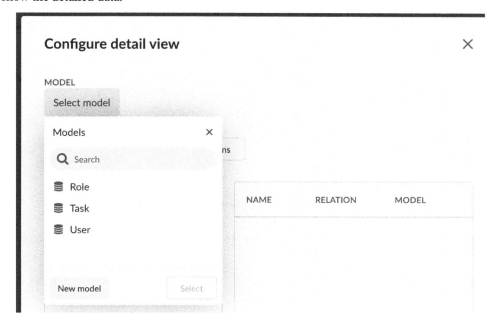

Figure 9.10 – The wizard where we can select our model

Here, we'll keep **2 columns** selected and after this, we will add all the properties to our detail view except for **Id**, by clicking on the + button to the right of each property:

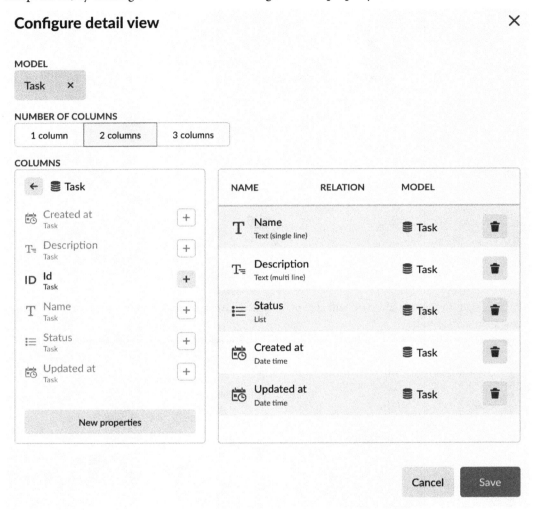

Figure 9.11 – The selected properties we want to show in our detail view

We have now finished selecting the properties for our view, so let's click on the **Save** button. The detail view is now shown in the dialog (*Figure 9.12*):

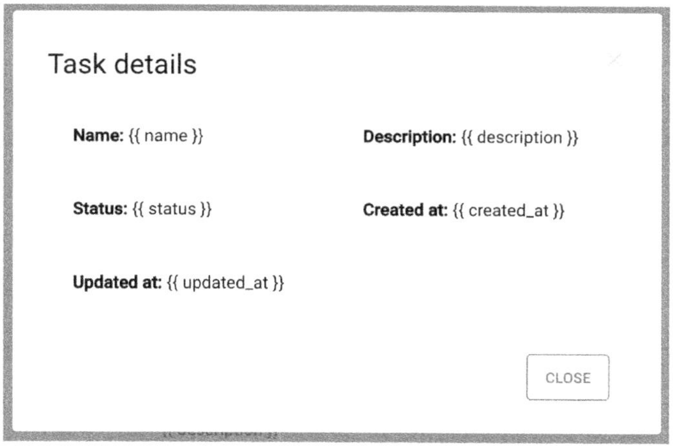

Figure 9.12 – The dialog with task details

Now, we're all finished with setting up our dialog, so let's hide it by clicking on the **Toggle visibility** button again, as shown in *Figure 9.6*.

We want to show our dialog by clicking on a button in the data table. To do this, we will search for a data table column, and then we'll drag it to the right-hand side of the data table:

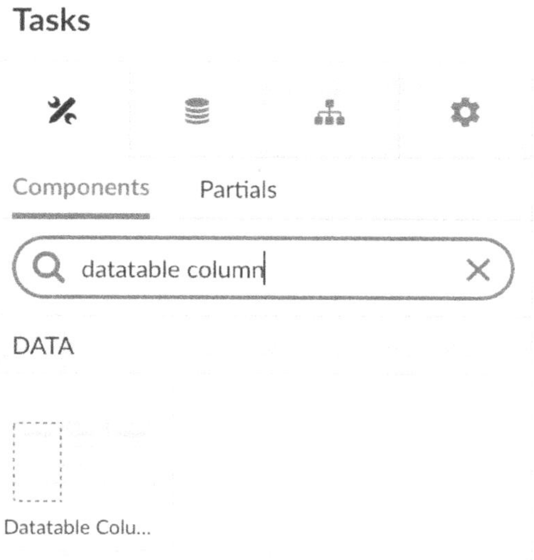

Figure 9.13 – The data table column component

When adding the data table column component, you will see a wizard where you can select a property. Here, we'll click the **Add without configuration** button, which will our new column in the data table (*Figure 9.14*):

Name	Description	Status	Select property
{{ name }}	{{ description }}	{{ status }}	Select property
{{ name }}	{{ description }}	{{ status }}	Select property
{{ name }}	{{ description }}	{{ status }}	Select property
{{ name }}	{{ description }}	{{ status }}	Select property
{{ name }}	{{ description }}	{{ status }}	Select property
{{ name }}	{{ description }}	{{ status }}	Select property

Figure 9.14 – The new data table column

Let's search for a button in our component set on the left and drag it onto our new data table column (*Figure 9.15*):

Name	Description	Status	
{{ name }}	{{ description }}	{{ status }}	BUTTON
{{ name }}	{{ description }}	{{ status }}	BUTTON
{{ name }}	{{ description }}	{{ status }}	BUTTON
{{ name }}	{{ description }}	{{ status }}	BUTTON

Figure 9.15 – The button in the new data table column

Now, let's select the button and change its appearance and also make the dialog visible when clicked. We'll remove the text and change the icon to **Visibility**, or any icon that you like (*Figure 9.16*):

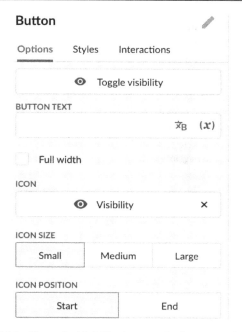

Figure 9.16 – Show the Visibility icon on the button without text

Now let's select the **Interactions** tab for the button and create an interaction that will show the dialog when a user clicks on the button. So when I *click* on the button, I want to *show* the **ViewDialog** component:

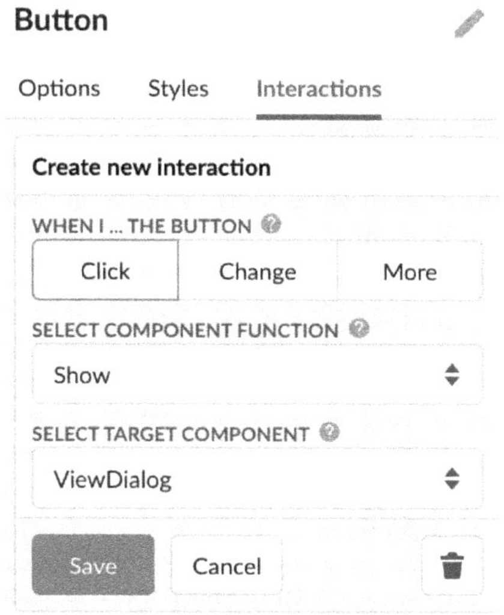

Figure 9.17 – Creating the interaction to show our View details dialog

The next interaction that we will make is for setting the current record for `DataContainer` in the dialog, so our dialog will know which data to use. To do this, we will need to select the `SetCurrentRecord` interaction in our list, which will set the correct record in our dialog. At the time of writing, there are currently two `SetCurrentRecord` interactions available. Make sure to select the second item in the list. After that, click on the database icon and select the ID property. This will tell the `SetCurrentRecord` interaction that it needs to use the task's ID to use the correct record in `DataContainer`. So, set **SELECT OUTPUT COMPONENT** to `DataContainer`.

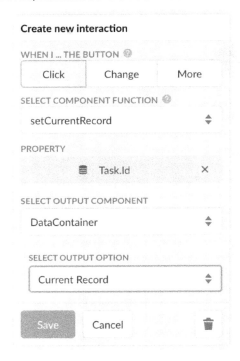

Figure 9.18 – Setting the current record of DataContainer in the dialog

Now let's compile our page and test what we've built. You should now be able to see the details of a selected item in your list, which are shown in the created dialog.

Edit tasks

Now we've made our first part in viewing our records, we would also like to be able to change the data. Therefore, we need to create an **Edit** form.

Let's get started by creating our second dialog on our **Tasks** page. Again, search for a dialog component (*see Figure 9.4*) and drag it below the first dialog. Now, rename it to `EditDialog`.

Your page should look like this now:

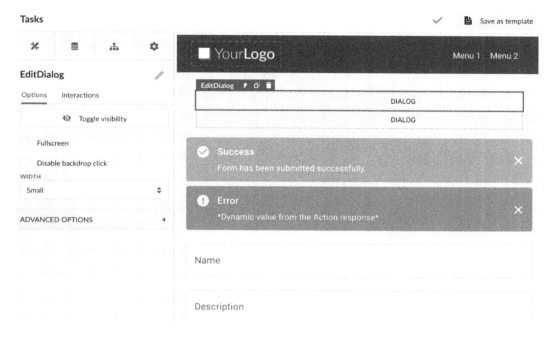

Figure 9.19 – EditDialog on the page

Now, we want to change the look and feel of the dialog we've just created. In order to be able to do this, select the dialog and click on **Toggle Visibility** (see *Figure 9.6*). Our dialog will pop up, and we can change the content (see *Figure 9.7*).

For the basic setup of our dialog, change the word **Dialog** to Edit task, remove the **SUBMIT** button, and remove the text block. Your page should look like this:

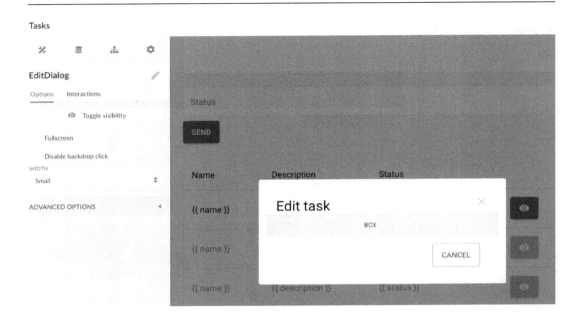

Figure 9.20 – The first step of setting up our Edit task dialog

Now, we'll search for the **Update form** component, and we will drag it in the box:

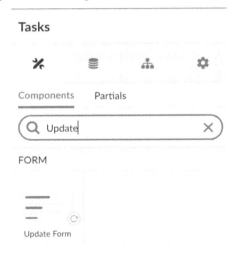

Figure 9.21 – The Update Form component

After dragging it in the box, a wizard will appear, in which we can select the model for which we want to edit the data:

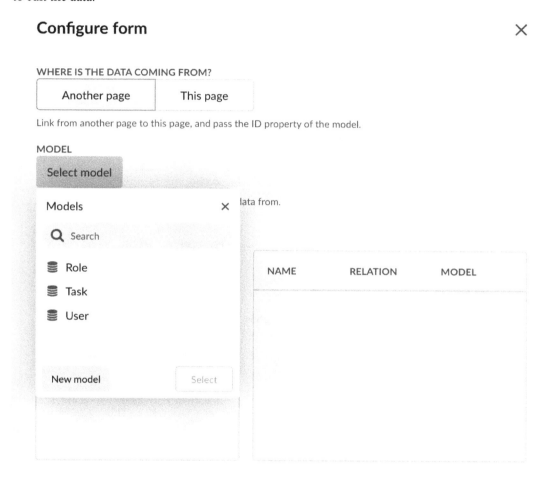

Figure 9.22 – The wizard where we can select our model

Select the **Task** model and add the properties you want to edit in the form by clicking on the + button. Also, make sure that you have selected the **This page** option, and in the dropdown, select **DataTable**:

Figure 9.23 – The final step in the Update form wizard

Now, let's click on the **Save** button and our updated form, including the action, will be created.

Let's change the caption of the **Send** button to **Save** and drag this button next to our **Cancel** button. Our buttons are now outside our form. To make sure we can submit and save our data, we should select the **Box** component in which our buttons are placed. To select this **Box** component, simply click on the **Cancel** or **Save** button and click on the upward arrow icon on the selected button:

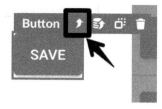

Figure 9.24 – The option to go to the parent component in the component tree

Now, let's drag this box component within our Form component. Your dialog should look like this now:

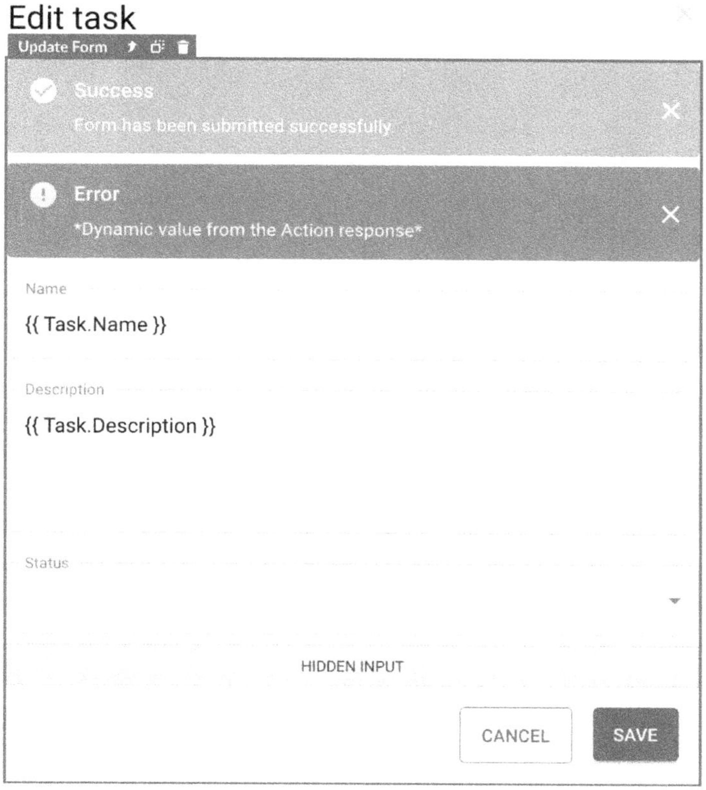

Edit task

Update Form

Success
Form has been submitted successfully ×

Error
Dynamic value from the Action response ×

Name
{{ Task.Name }}

Description
{{ Task.Description }}

Status

HIDDEN INPUT

CANCEL   SAVE

Figure 9.25 – The final Edit dialog

Now, we're all finished with setting up our dialog, so let's hide it again by clicking on the **Toggle visibility** button again, as shown in *Figure 9.6*.

We want to show our dialog by clicking on a button in the data table. To do this, we will first search for a data column (see *Figure 9.13*), and we'll drag it to the right-hand side of the data table.

When adding the **Datatable Column** component, you will see a wizard where you can select a property. Here, we'll click the **Add without configuration** button, showing our new column in the data table. Let's add a button as we did before on our new data column (see *Figure 9.15*).

Now, let's select the button and change its appearance and also make the dialog appear when clicked. We'll remove the text and change the icon to **Edit**, or any other icon that you like:

Figure 9.26 – The configuration of the Edit button

Now, let's select the **Interactions** tab and create an interaction that will show the **Edit** dialog when a user clicks on the button. So, when I *click* the button, I want to *show* the **Edit** dialog. For information about the configuration, please refer to *Figure 9.17*.

The next interaction that we will do is to set the current record for our data container in the dialog. To do this, we will need to select the `SetCurrentRecord` interaction in our list, as used before (see *Figure 9.18*).

The last interaction we will create is an interaction to refetch our data in our data table and hide our **Edit** dialog. These should be added to the **Update Form** component. Try to create these for yourself.

Now, let's compile our page and test what we've built. You should now be able to edit the details of a selected item in your list.

Delete tasks

Now we've completed our first functions of viewing and editing our records, we would also like to be able to delete the data. Therefore, we need to create a **Delete** form.

We want to show a confirmation dialog by clicking on a **Delete** button in the data table. To do this, we will first search for a data column (see *Figure 9.13*), and we'll drag it to the rightmost part of the data table. When adding the **Datatable column** component, you will see a wizard where you can select a property.

Here, we'll click the **Add without configuration button**, showing our new column in the data table. Let's search for a `Delete record` button and drag it in our new data column (see *Figure 9.15*):

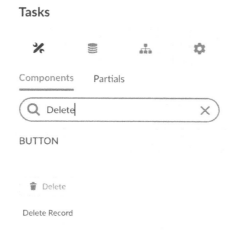

Figure 9.27 – The Delete button to delete a record

After dragging this button, a wizard will appear, where we can select our **Task** model (see *Figure 9.28*). After selecting this model, click on **Save**:

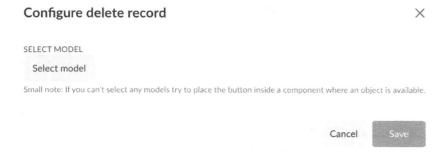

Figure 9.28 – Configuring the delete record button

The dialog will be created automatically for us, including most of the interactions. The only extra interaction we need to configure is a refetch of our data table. Try to configure this yourself, within the **Delete** button component of the dialog.

Also, let's remove the `Delete` text from our **Delete** button. Your data table should look like *Figure 9.29*:

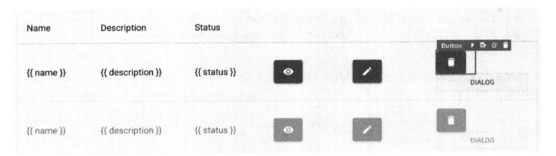

Figure 9.29 – The data table containing the new Delete button

Now, let's compile our page and test what we've built. You should now be able to delete a selected item from your list.

Summary

In this chapter, we've learned how we can view the details of our records by creating a dialog and an extra button to view the dialog itself. We've also learned how we can edit data for a specific record using an **Update Form** component in a dialog. Next to this, we've learned how to delete data using a confirmation dialog. Also, we've added extra interactions on the different dialog.

Questions

1. What component is used to easily create a view of your data for a single record?

2. Which form do you use to update your data in a model?

3. What does the **Delete** button do?

Answers

1. To show data from a single record, the easiest way to do it is to use the Detail view component. It allows you to use a single record from a model and show the data it holds with a lot of effort.

2. The Update Form is used for this. It automatically creates an action for you that handles updating the data and renders all the inputs based on the properties that you select.

3. The **Delete** button creates an action that deletes data from the model.

10

Case Management Application

In this chapter, we'll focus on creating the functionality of a case management application. A case management application is an application that allows you to manage a case for a customer, such as an insurance company or law firm, among many other use cases. A case holds information about a specific case, where you can track the status of the case and save notes related to it. This should further familiarize you with all the basics that you've been taught in the previous chapters. The application will make use of a data model, the page builder, actions, and interactions for our logic. We will also be using authentication in this application.

This chapter is divided into the following parts:

- Using the login template
- Creating our data model
- Setting up the Back Office

After this chapter, you will have a better understanding of the process of creating an application, setting up authentication, creating advanced pages with actions, and tracking data changes in your application.

Using the login template

By using a login template, we can easily set up the basics of our application before developing them further later. Let's create our new application in **My Betty Blocks** (`https://my.bettyblocks.com`) by clicking on the applications button, represented by the nine dots in the sidebar (see *Figure 10.1*).

Figure 10.1 – Selecting the applications button in My Betty Blocks

In the top bar, you will see a **Create application** button, which we select to create a new application.

Figure 10.2 – The button for creating a new application

Now, search for the login template and select this item:

Figure 10.3 – The login template in My Betty Blocks

In our new application, the basics for our login process have already been made for us. This is done by using an authentication profile. An authentication profile defines which model or models your application will use to validate any users attempting to access a protected (authenticated) page. In the majority of cases, the user or web user model is used as it contains information unique to individual users, but this can be customized when configuring a new authentication profile.

Our app already has a login page and a configured authentication profile. Web users attempting to log in will be checked to ensure their unique identifiers (generally an email address and password) match your authentication model and that a user record exists with matching credentials.

You can find your current authentication profile by clicking on the **Authentication profiles** option in the builder bar:

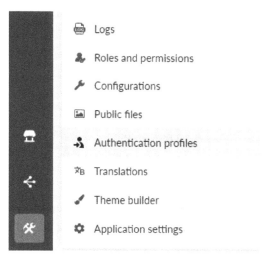

Figure 10.4 – The menu option for authentication profiles

From this overview, you can configure new authentication profiles or change the settings of an existing profile.

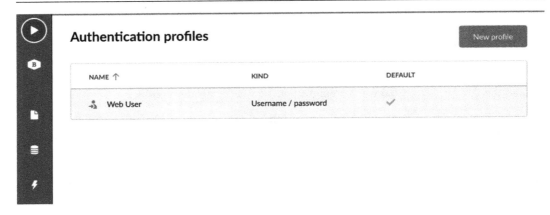

Figure 10.5 – The overview of our current authentication profiles

Also, extra pages are created for us. Let's take a look at them:

- **Home**: This is our starting point in the application
- **My account**: Here you can change settings for your account
- **Create account**: Here we can create a new account
- **Login**: This is our actual login page
- **Password reset request**: Here we can make a password reset request
- **Password reset**: This is the page where we actually reset the password
- **Web user management**: This is a Back Office page where you can manage the current web users in your application

Make sure to check that the live versions of your pages work. If by chance they don't work, please recompile the page by clicking on the play button. Whenever your page doesn't work or doesn't look the way you expected, always try to recompile it first by opening the page and pressing the play button.

Now that we've created this application, we need to create a web user and test the login flow. In order to do this, we will first open our application on the /login page. When opening the page, you might encounter an error saying *no login page is yet set for the current authentication profile*. When opening the authentication profile, you can set this with the following dropdown:

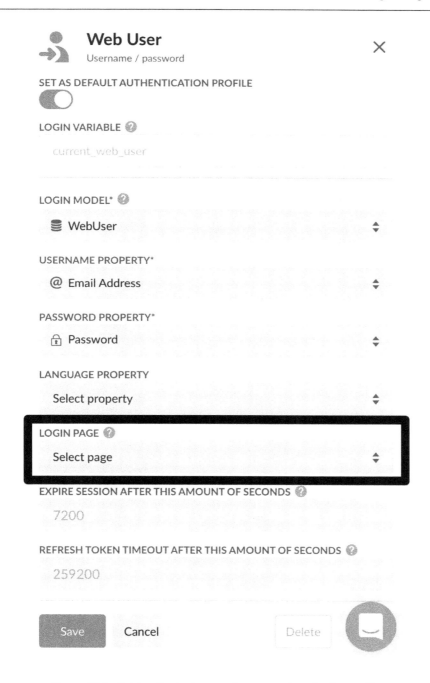

Figure 10.6 – Setting the login page for your authentication profile

After opening the login page, we need to select the **REGISTER** option:

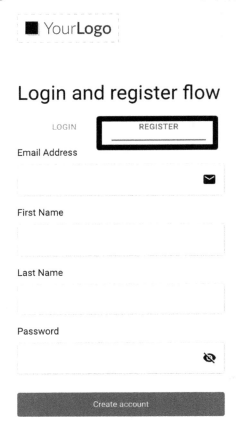

Figure 10.7 – Registering a new web user

Now fill in all the fields and click on the **Create account** button. Your new user is now created and you can test the login flow. Let's click on the **Login** button and fill in your newly generated user account.

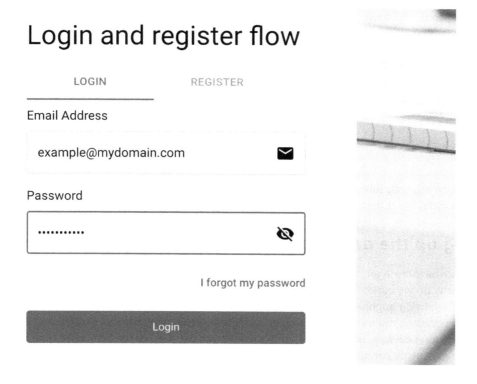

Figure 10.8 – Filling in your login credentials

After filling in the credential data and logging in, you should be redirected to the home page, which for now is just a simple empty page.

Figure 10.9 – A simple home page

Now let's open the **Web user management** page. Your new account should be shown here.

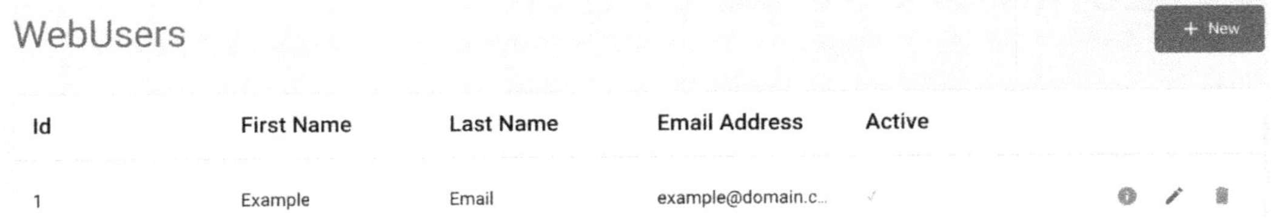

WebUsers								+ New
Id	First Name	Last Name	Email Address	Active				
1	Example	Email	example@domain.c..	✓		ⓘ	✏	🗑

Figure 10.10 – Overview of current web users

Now we've created our first user and are able to log in, it's time to get started on the data model to allow us to insert some data into our application.

Setting up the data model

Now that we have our login functionality set up for us, we can start building our data model straight away. So, why do we build our data model first, instead of building our pages first and adding our data model later, you might ask yourself? Both are valid ways of doing it, but if you're building a more complex application, you want to get your data model right from the start. That's why I prefer to create my initial data model first, so I have everything ready from the start. In the data model you also have the option to directly configure all your settings, such as required or default value which you don't have when you're quickly adding your properties from the page builder. You can always do that later, of course, but I find the risk of forgetting something and creating a bug in your application is too high. So that's why, in this book, we always start with the data model first: it's a preference, not mandatory.

This data model will be larger than our previous ones since this application will have a lot more functionality. Often, you might only make a part of your data model first because you only want to build the first part of your application, but since this application isn't that large, we can do everything in one go. So let's get started!

Click on the data model icon in your builder menu – it's the database icon that should light up in purple when you hover over it. When the sliding pane opens, click on the plus button to add a new model (see *Figure 10.11*).

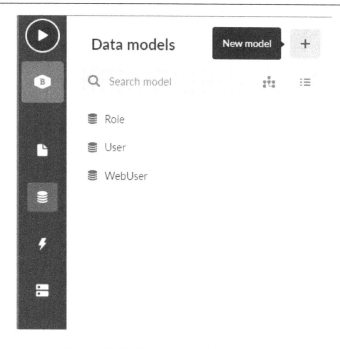

Figure 10.11 – The Data models pane open

The case model

The first model we'll add is the case model. This will be the main model for our application since it will contain all information related to the case. You should have the **Create new model** dialog open right now – let's enter the name Case here and click on the **Create model** button to create a new model. This should take you to the **Properties** overview page for your model (see *Figure 10.12*).

← **Case**

Properties   Relations   Permissions   Validations   Settings

Properties (3)   Q Search   Quick add properties   New property

NAME ↑		DATABASE NAME	DEFAULT VALUE
Created at Date time		created_at	
Id Serial	⌐ ↕	id	
Updated at Date time		updated_at	

Figure 10.12 – The properties overview for your case model

Now we can create new models in two ways: using the **Quick add properties** option or the **New property** button. Since we also want to set some options for our properties, we'll use the **New property** button. So let's add our first **Name** property, setting it to the **Text (single line)** type with the **IS REQUIRED** option toggled on, so the user always has to fill in this field before they can save their record in the model. Then, hit the **Save** button. Your page should look like *Figure 10.13*.

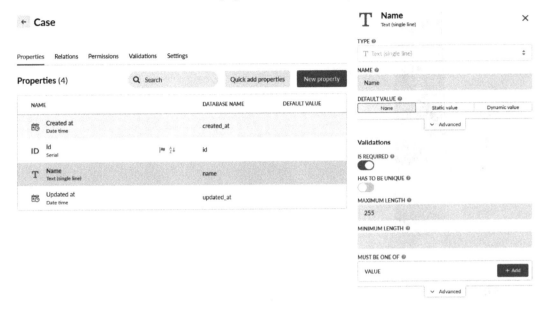

Figure 10.13 – The Name property added to your model

This is our first property. We'll now add some more properties, going over the details of unfamiliar new ones and the specific options that we'll use for each.

Next, we'll add a **Description** property of the **Text (multiline)** type. This type of property is similar to **Text (single line)** with the difference that the single-line type only accepts a maximum of 255 characters and can't save any additional lines of text, while the multiline type accepts a huge amount of characters and can also handle new lines. You can use multiline for large bodies of text, but also to save HTML, for example. Don't forget to press the **New property** button first to allow you to add the **Description** property and note that we won't set any options for this one.

Next, we'll add the **Status** property. This will be of the **List** type, and as you may remember from before, we need to set some values for list properties. This is because a list property offers the user some predefined values to choose from, and only those. So let's add the following values:

- New
- In Progress
- Completed

These should suffice for us to set and update the status for our cases. The default value is important here because we want every case to start off with the status of **New** by default. To do this, select the **Static value** option and then select the **New** value. Let's also toggle the **Validations IS REQUIRED** setting to on, so it will always have a value.

At the time of writing this chapter, the color options for list values only worked in the classic Back Office, so we will not set these here. This might have changed by the time you are reading this, so you can always set them and see whether they appear later on in the page builder. Your list property settings should look like *Figure 10.14*.

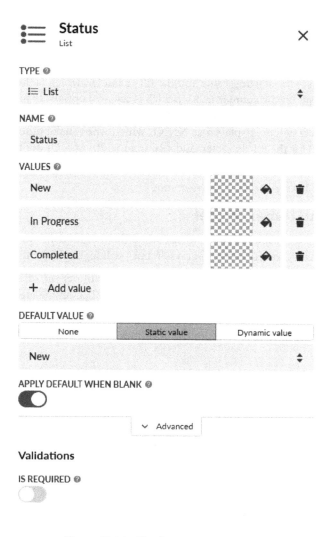

Figure 10.14 – The Status property set up

The last property we'll add is a completely new one that we haven't seen before. It's the **Auto Increment** property. This property automatically increases in value every time you create a new record. That makes this property quite powerful and means you don't have to worry about manually incrementing a value to keep track of record creation, which, while possible, is very error-prone, so I'd recommend doing this instead by creating this property. At the time of writing, there was only one **Auto Increment** property, but there will soon be a new one since the old one isn't compatible with the newer versions of the platform. The old one should have the name **Classic** beside it, while the new one should have the usual **Auto Increment** name. Since there are no expressions (they're like formulas) available anymore like there used to be in the classic version of the platform, the new one that will be released has some specific features, which we mostly used when people created this property in the past.

Choose the **Auto Increment** type and name it Case number. As you can see, the format for our number has already been set. We'll stick with the default format for now, which gives us the year number with a number beside it that automatically increases every time we create a new record. We also have the options to set the starting value and the fill character, which is the character that will be used, for example, if you set the counter width (which is the next option) to four: the first number will be **0001** then, with 0s being used as fill characters. If you changed the fill character to N, for example, the first iteration will be displayed as **NNN1**, which won't make much sense. We'll use 1 as our starting number, 0 for the fill character, and a counter width of 4. Under the format you see a link to an explanation of how you can change the format, which you can read if you are interested.

So that's our case model. Let's move on to the next one.

Setting up the customer model

Our next model is the customer model. Since we just set up the case model, I won't go into too much detail on each property here to avoid repeating myself, but I will highlight anything new.

The customer model will hold all the information about the customer on which the given case is based. Click on the data model icon in the builder bar and then click on the create new model icon there. Let's name the model Customer and create it.

We're going to add the following properties to the customer model, providing the property name, type, and options to set:

- First name – **Text (single line)** – **Is required**
- Last name – **Text (single line)** – **Is required**
- Street – **Text (single line)** – **Is required**
- House number – **Text (single line)** – **Is required**
- Postcode – **Text (single line)** – **Is required**
- City – **Text (single line)** – **Is required**

We've set the house number property to single-line text instead of a number. You might be wondering why, since house numbers are numbers. This is true, but in a lot of countries there are letter additions to the numbers (if that's not the case for you, then you can of course change it to the number type, but text gives you more flexibility as the number type won't allow anything other than numbers).

We still need to add two more, namely the email address and phone number properties. Let's start with the email address. We'll use the `email address` property type for this one. It's a specific property type for email addresses, that's basically a number, but it validates that the input value is an email address by checking for an @ symbol and a dot in it. It also checks that the extension at the end is correct as well, so users will always enter a valid email address. We'll name it `Email address` and set **Is required** to on, and we'll also set the **has to be unique** option to on. This makes sure that we don't have any customers in our model with duplicate email addresses.

Next up is the phone number property. This property doesn't do any validation checks that phone numbers have been entered properly since different countries have different ways of writing phone numbers. It's possible to set this up in the phone number property itself, but we are not going to do that now. The main reason is that we also have a phone number input in the page builder, which works very well together with this one, so it will be easier to set it up later in the page builder instead. So, let's create this one and make it required like the previous ones.

Now that we have all of the properties set up for our customer model, our data model should look like *Figure 10.15*.

NAME		DATABASE NAME	DEFAULT VALUE
T	City Text (single line)	city	
📅	Created at Date time	created_at	
@	Email Address Email address	email_address	
T	First name Text (single line)	first_name	
T	House number Text (single line)	house_number	
ID	Id Serial	id	
T	Last name Text (single line)	last_name	
📞	Phone Number Phone number	phone_number	
T	Postcode Text (single line)	postcode	
T	Street Text (single line)	street	
📅	Updated at Date time	updated_at	

Figure 10.15 – The customer model properties

Now let's set up our next model, the case note model.

Setting up the case note model

This is a small model since it will only be holding the notes added by the user for a specific case. Let's go to the builder bar, click on the data model icon, and then click on the new model icon. Let's name this model `CaseNote` (you could also name it `Case Note` if you prefer – the first is a bit more in line with the traditional programmer style using a format called camel case, but both are correct). After you've saved it, let's add the properties. It will only have two properties:

- `Title` – **Text (single line)** – **Is required**
- `Description` – **Text (multiline)** – **Is required**

So, that's it for the properties. Are we done now? No, not really – in our work on models so far, we haven't yet gone into relationships between models, so let's add some relationships to this model now. There are two ways of adding relationships to your model: one is from the model itself by going to the **Relations** tab there. The other is in the schema view, which is a graphical representation of the models. We'll use the latter here since the visual element will make it a bit easier. Both will have exactly the same outcome.

In order to get to the model overview, we'll need to click on the data model in the builder bar and then under the **New model** button, you'll see two buttons. We need to click on the left one that looks like a flow chart. This will send you to the schema view, where you should see the following models: **User**, **Role**, **Case**, **Customer**, and **CaseNote**. If you don't see these, click on one of the models, hold down your mouse button, and drag it away a bit from where it was initially placed. Do this for all the models until you see all of them. You should end up with something like *Figure 10.16* – it doesn't have to look exactly the same, as long as all the models are present.

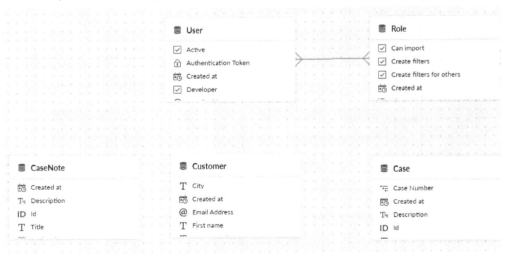

Figure 10.16 – The models shown in the schema view

So we'll need two relationships for our **CaseNote** model. The first one is a relationship with the user, so we can store the user who created the note. The second one is a relationship between **CaseNote** and **Case** so we can see for which case the note was made.

In order to create a relationship in the schema view, we need to drag an arrow from one model to another. Let's start with **CaseNote** and **User**. Hover over the **CaseNote** model with your mouse and then you'll see an arrow appear (*Figure 10.17*).

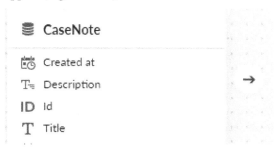

Figure 10.17 – The arrow next to the CaseNote model

Click and drag from the arrow to the web user model. A dialog box will appear (*Figure 10.18*). Here you'll see three options as follows:

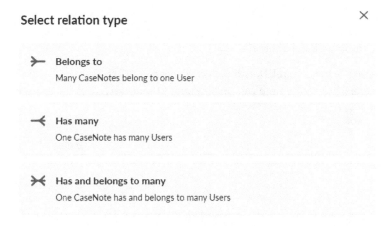

Figure 10.18 – The Select relation type dialog box

These three options each create a very specific relationship between the models. At first, I can imagine that these might be a bit overwhelming. You'll understand them better once you've used them a few times. The descriptions under each of them should help a bit with understanding what they do.

Belongs to (Many CaseNotes belong to one User) is used if your user might create more than one note in your application belonging to them. This makes sense, because it wouldn't be helpful if a user could only write one case note for a case. It specifically means that a case note can only have one user on it, which is also what we want because that way we can track exactly which user has written each note.

Has many (One CaseNote has many Users) is basically the opposite of **Belongs to**. This is not what we want because our case notes should only have one user, and it will also mean that a user can only have one case note.

Has and belongs to many (One CaseNote has and belongs to many Users) means that you can have many case notes belonging to many users. So a case note might have multiple users, and a user might have multiple case notes. This might make sense in some cases, but it will also make it a lot more complicated, so it's not what we want for this use case.

We'll go with **Belongs to** so we can track which user has written which note, and also allow a user to write more than one note. So let's click the **Belongs to** option now. You'll see that a line is drawn between the two models and is clickable, so you can view it and delete it. Note that you can't change relationships; you can only delete them and create a new one. Furthermore, you can create multiple relationships between models if needed.

Alright, so now we still have to create a relationship between the **CaseNote** and **Case** models. Let's do the same again as before: hover over the **CaseNote** model, then click and drag the arrow to the **Case** model. The same three options should appear again.

In this case, a case note will belong to one case, while a case can have multiple case notes. So which relationship do you think we'll need there?

If you've chosen the has belongs to relationship, you are correct. It's the same as with the user: a case can have multiple notes, but a case note can only belong to one case. Your data model should now look something like *Figure 10.19*.

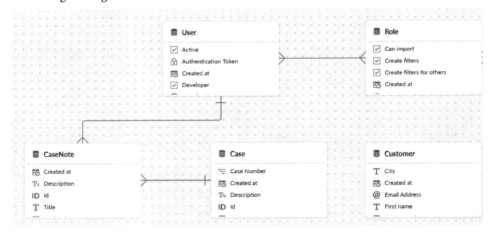

Figure 10.19 – CaseNote with its relationships

So now we have our first relationships set up. Let's continue our work in the next section by adding the last two models and their relationships.

The last two models

In this section, we'll add the last two models. Both are small models and, like **CaseNote**, mostly support the case model. The **CaseHistory** model will allow you to store previous status changes for a case, so you have a simple history of what the previous status was and when it was changed. The **CaseFile** model will allow you to upload files to the case, but instead of allowing only one file to be uploaded, this model will allow the user to add multiple files to a case. In the following list, you'll see the name of the models with their properties: go ahead and create them as outlined in the previous sections. After that, we'll add some relationships between them and the other models:

- CaseHistory:
    - Old value – Text (multiline)
    - New value – Text (multiline)
- CaseFile:
    - File – File – Is required
    - File name – Text (single line) – Is required

Now that you've created all the models, let's add the relationships. Since those might still be hard to understand for citizen developers, we'll go through them.

The CaseHistory model has two relationships. The first is a relationship with the Case model, which allows you to track the changes in this model and see them in your case. Since the case will have multiple histories, the relationship will be the case model has many case histories. So, the easiest way to create the correct relationship is to drag an arrow from the Case model to the CaseHistory model. Do this just like we did before, hovering over the Case model, grabbing the arrow, and dragging it to the CaseHistory model. You'll get three options again – the second option reads **One Case has many CaseHistories**, which is exactly what we want, so choose that one.

Then we also want to track which specific users have made what changes to the case and track this in the history. For that, we'll need a relationship between the CaseHistory model and the Web User model. A case history will have one user, so drag the arrow from the CaseHistory model to the Web User model and choose the **Belongs to** relationship.

We still need to add relationships to the CaseFile model. We will have just one relationship between the Case model and the CaseFile model. Drag the arrow from the Case model to the CaseFile model. Each case will have many case files. This is exactly what's written under the second option, **Has many**, so let's choose that one. Your data model should now look something like *Figure 10.20*.

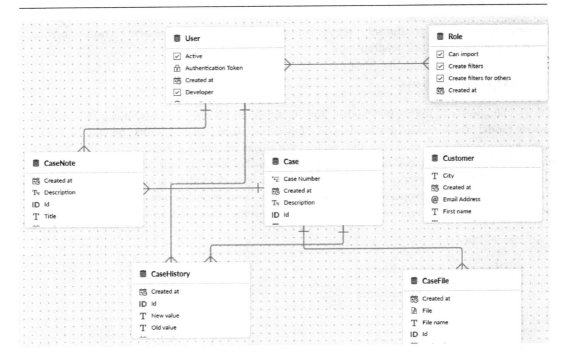

Figure 10.20 – The data model with relationships

Now the only thing left to do is add the last relationships between the customer and the case. A case always has one customer, while a customer can have multiple cases, so let's drag an arrow from the Case model to the Customer model. Which model will you choose?

Belongs to is the correct one. Many cases belong to one customer. And that's it for now. We've got our data model set up and we can get started on creating our Back Office, so we can interact with some parts of our data model.

Setting up the Back Office

We want to make data management as straightforward as possible, and that's where the Back Office comes into play. This part of your application is only accessible to you and your fellow employees, but not external web users.

It's very simple to create a Back Office application by using the Back Office page template. Let's create this for our user management in the page builder, so that we can manage our users easily from the Back Office.

A small note here: at the time of writing this chapter, the Back Office template is being rebuilt, so it might look different and have more options by the time you read this book, but the basics should remain the same.

Setting up the user management

We'll start with creating a new page by going to the builder bar and clicking on the page builder icon, and then we'll click on the create a new page icon, and we'll select the **Back office** page template in order to create our first Back Office page.

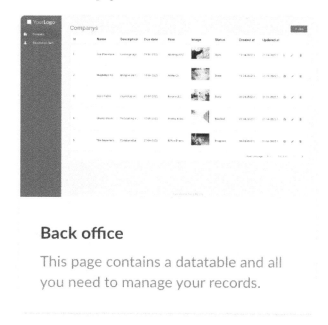

Figure 10.21 – The Back office page template

We'll name the page User management. Next, we need to choose a type for our page – we want to set it to **Authenticated** and then choose the **Betty Blocks account** profile. This is the default profile that comes with the platform and uses internal authentication. This means that it uses the same users as those that you can add and manage through My Betty Blocks.

Now click on the **Create page** button. Next, we'll get a page where we have to select our partials (we'll cover partials in more detail later in this chapter). Most likely, you will already have a footer partial but not a menu partial. If this is not the case, click on the **Auto generate** button to create the partial. Once that is done, click on the **Next** button to go to the next step.

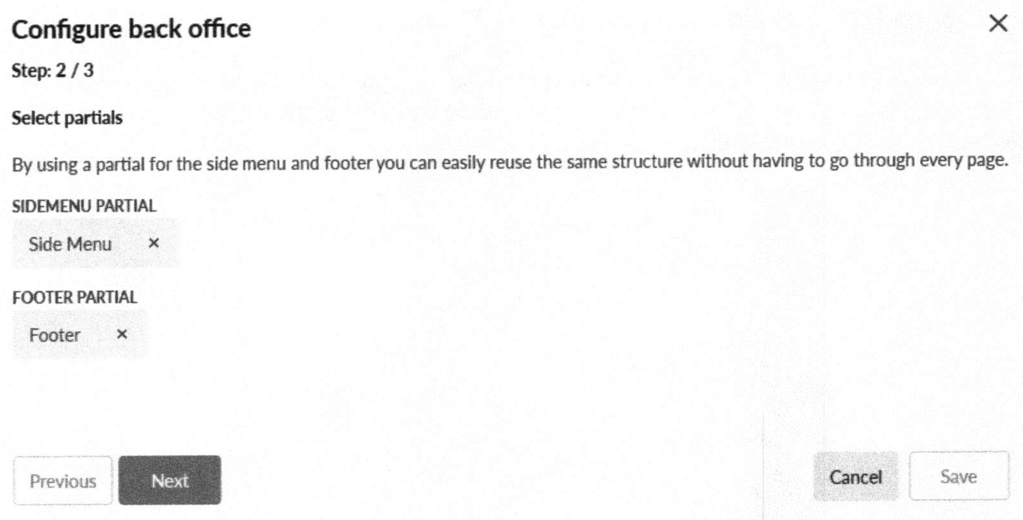

Configure back office   ✕

Step: 2 / 3

Select partials

By using a partial for the side menu and footer you can easily reuse the same structure without having to go through every page.

SIDEMENU PARTIAL

| Side Menu   ✕ |

FOOTER PARTIAL

| Footer   ✕ |

| Previous | Next | | Cancel | Save |

Figure 10.22 – Both partials have been generated for you here

The last step is to choose which model we want to use for our Back Office. This will be our **WebUser** model, which holds all the models for the frontend part of the application. Click on the **Select model** button and chose the **WebUser** model. Next, we'll need to select the properties that we want to add. Add the following properties by clicking on the plus button:

- **First Name**
- **Last Name**
- **Email Address**
- **Active**

These models will show up in our data table in the Back Office. The others will be added automatically to the create, update, and view parts of our Back Office. Lastly, let's hit the **Save** button to create our Back Office page.

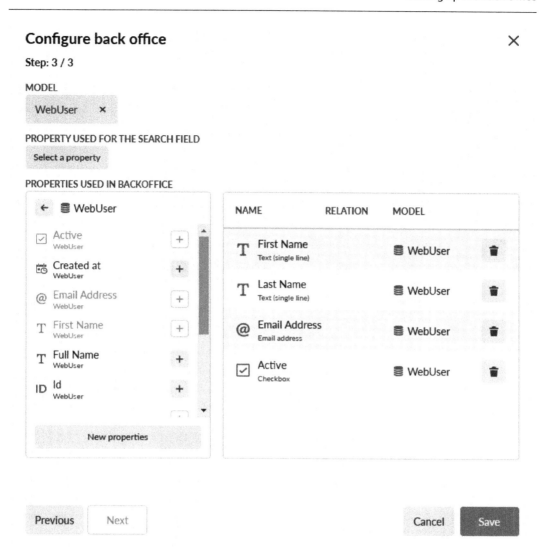

Figure 10.23 – Selecting the WebUser model while creating the Back Office page

After clicking this button, our Back Office page will be generated. Now let's click on the play button on top of the builder bar to compile the page, and when complete, we can now open it. Click around a little to get more familiar with it.

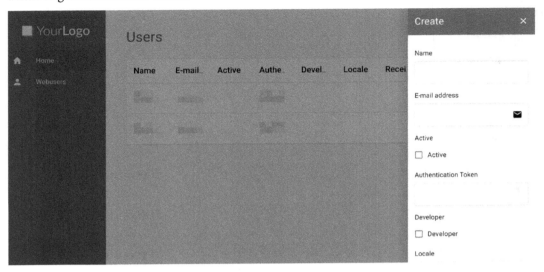

Figure 10.24 – Our newly created user management Back Office page

Next, we'll create the same page for our cases. Why? Because you might want to edit or delete some cases as an admin with the ability to override your users in some specific cases. Follow the exact same steps as previously to create a Back Office page for your Case model. Don't forget to compile the page by clicking on the play button in the builder menu.

Now we need to add these pages to the menu. Open either of the Back Office pages we created before. Now click on the menu in the page builder – there should be a green border around the menu component. The reason it is green is that the component you just clicked is a partial. We've mentioned these before, but now we're going to edit this one, so let's go over what a partial exactly is.

Changing your partial

A partial is a component just like the others, but with one big difference: this component stays the same across every page you add it to. So, if you make a change to this partial, that change will apply on all pages. You will notice that if you try to drag a component inside the partial you won't succeed, because the part itself needs to be edited in the partial editor to make it look the same on all pages.

You'll notice on the left side that there is a button called **Edit partial**. With this button, we can edit our partial, so let's click on it. Some changes might have been made here by the time you read this – there might be a button that says *add a new menu item* or something along those lines. You can use that as well, but if you follow along with me for the first change, you'll learn how to make changes manually as well.

You might see a huge menu now that covers the whole page. This is only because there is no specific column formatting applied to make it smaller – don't worry about this, it will look normal again when it's on the page. Click on **First list** item and on the left side in the options, change the primary text to Web Users. Then below that, click on the blue **Select page** icon and choose the web user Back Office page. If you like, you can even select a nice icon that will fit the purpose of the page.

Then click on **Second list item** and rename it to Cases. Change the page and icon as well. Now, at the top, you'll see an arrow to click on that will bring you back to your page. After clicking it, your menu should look like *Figure 10.25*.

Figure 10.25 – The changed menu

Click on the play button in the builder menu and try out your page with the new menu. Now that you have a working Back Office, it's time to build the rest of your application. We'll continue with this in the next chapter, building the remaining pages.

Summary

In this chapter, we started on our case management application. We first learned about how to set up and use the login template. After that, we set up our data model. We covered some new properties and their options, including email address, auto increment, and phone number. We also set up some relationships between the models so we can use data from other models later in our application, such as to set notes on a case or a user on our note changes. Once our data model was set up, we created a Back Office that allowed us to make quick changes to records in our data model, or create entirely new records. The Back Office is specifically designed for admins to enable them to make these changes.

In the next chapter, we'll finish the application by creating the pages and diving deeper into the actions for the first time.

Questions

1. What do you need to create a login?
2. Which type of property do you use to create a status?
3. Which three types of relationships does Betty Blocks support?
4. Which page template is used for creating a page to manage your data easily?

Answers

1. You'll need an authentication profile, which helps you to set up a login profile.

2. You use a list property. This property has multiple preset values that a user can choose from.

3. **Belongs to**, **Has many**, and **Has and belongs to many**.

4. The Back Office template. This template offers out-of-the-box functionality to make an administrator environment for properties.

11

Case Management – Pages and Actions

In this chapter, we'll focus on creating the functionality of the frontend of our case management application. This frontend will allow the end user to manage a case for a customer. This could be for an insurance company or law firm, for example, among many other use cases. Each entry in the data model will hold information about a specific case, where you can track the status of and save notes on the case. This should further familiarize you with all the basics that you've been taught in the previous chapters. It will be a frontend that will connect to our created data model, the page builder, actions, and interactions for our logic. We will also be using authentication in this frontend. In the previous chapter, we used the Betty Blocks account as the authentication. In this chapter, this will be the web user authentication profile.

This chapter covers the following topics:

- Creating an overview page
- Creating a new case
- Adding notes and files to the case
- Changing the status of the case
- Tracking the history of your case
- Showing the case details page

By the end of this chapter, you should have a better understanding of the process of creating a frontend, using authentication profiles on different pages, creating advanced pages with actions, and how you can track data changes in your application.

Creating an overview page

In the previous chapter, we set up authentication for our application. This will be used in our frontend. After we've logged in to our frontend, we want to be redirected to an overview page, from where we can manage our current cases.

For our overview page, we will be using a predefined template. By using a predefined template, you can quickly build a lot of the basic functionality you need in your application. These predefined templates already contain a lot of basic functionality, which will be further extended in the future as well. This will save you even more time later on.

Let's create a new page by clicking on the **New page** button, as shown in *Figure 11.1*.

Figure 11.1 – The button for creating a new page

After selecting this button, we will need to select our predefined template. For this template, we will be selecting **CRUD with slide-out panel**. See *Figure 11.2*.

CRUD is an abbreviation for **create**, **read**, **update**, and **delete**. So, using this predefined template will give you an out-of-the-box functionality for performing these tasks.

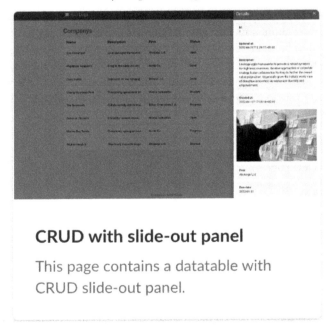

Figure 11.2 – Predefined template with a slide-out panel

Before using a preferred template, you can preview it by hovering over it and clicking on the **Preview** button. If you want to use the template, click the **Use** button.

CRUD with slide-out panel

In this ready to use Data Table, it is possible to create, display (read), update and delete records. These functionalities are shown in a slide-out panel.

Use   Preview

Figure 11.3 – Extra information on the predefined template with the option to use or preview it

After choosing the predefined template, we will give it the name `Overview`. Make sure to set the type of page to **Authenticated**, as we want to make sure the data is not publicly available. After this, we will link our web user profile to this page, making sure users first need to be logged in to perform other actions.

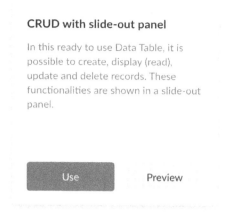

Page settings

PAGE NAME

Cases

PAGE TITLE

PATH

/cases

https://case-management-book.betty.app/ca...

SEO DESCRIPTION

The SEO description is a short snippet of text summarizing the content of the page which can be found by a search engine.

COMPONENT SET

Default

AUTHENTICATION PROFILE

Web User   ✕

Figure 11.4 – The setup of our new Overview page

Now, we're all finished, and we can click the **Create page** button. In the second step, we need to set our side menu and footer partials.

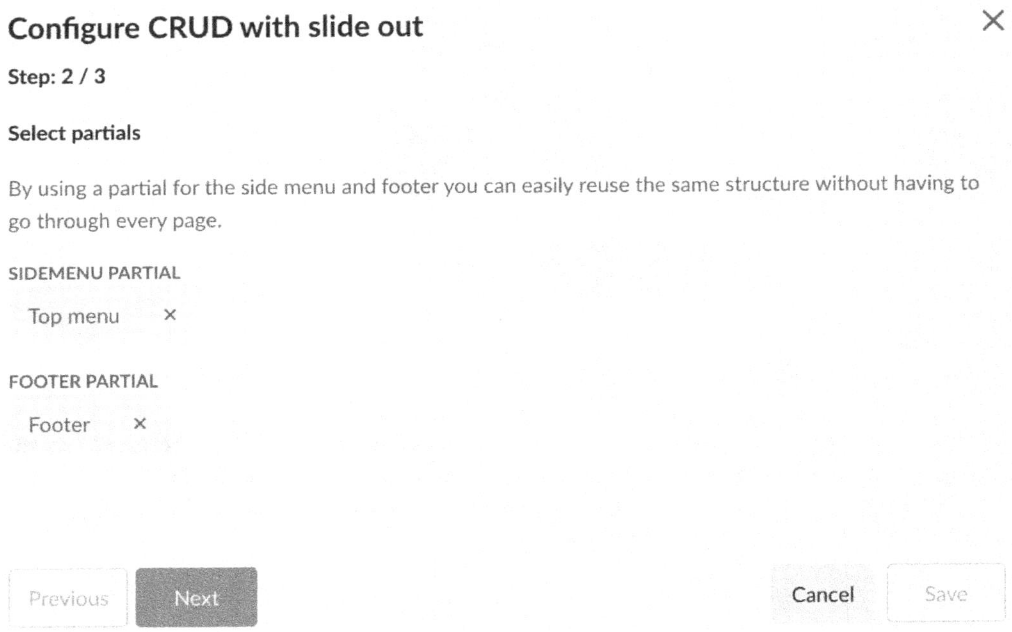

Configure CRUD with slide out   ✕

Step: 2 / 3

Select partials

By using a partial for the side menu and footer you can easily reuse the same structure without having to go through every page.

SIDEMENU PARTIAL

Top menu   ✕

FOOTER PARTIAL

Footer   ✕

Previous   Next   Cancel   Save

Figure 11.5 – Setting up the partials for our new page

Now that we've set up our partials, we can click on the **Next** button to move on to the final step. In the final step, we need to configure our linked model and the data we want to show in the different forms.

Let's select the **Case** model first. After selecting the model, we can select the properties that we want to use in our forms. Let's use the following properties for our page:

- **Name**
- **Description**
- **Created at**
- **Updated at**

We can select these properties by clicking on the + sign, which will add them to our form.

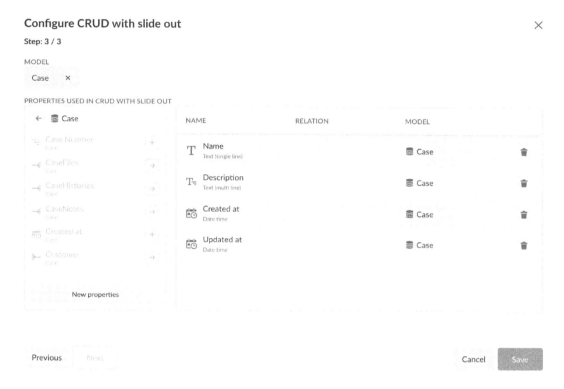

Figure 11.6 – The selected properties for our Case overview page

After selecting the properties, we can simply click on **Save** to create our new page. After creating the new page, you will see the following warning:

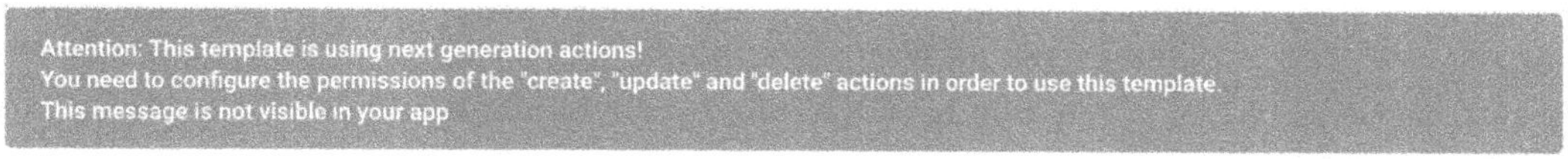

Attention: This template is using next generation actions!
You need to configure the permissions of the "create", "update" and "delete" actions in order to use this template.
This message is not visible in your app

Figure 11.7 – Warning about the action permissions

In order to be able to really start using our page, we first need to make sure all the permissions are set up correctly for the actions used on this page.

To be able to set the permissions, we need to look for the different forms in our component tree. Make sure you take the following steps for **Create form**, **Update form**, and **Delete Form**.

For this example, I will use the **Create form** component. When opening the component, you will find this component somewhere in the tree. After selecting the component in the component tree, click on the three dots and select **Options**.

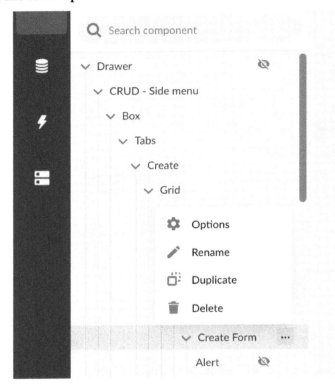

Figure 11.8 – The selected Create Form component with the Options setting

After clicking on the **Options** menu item, the options for this form will appear. You will see that an action is linked to this form. This is the action that will be executed after clicking the **Save** button. Just below the linked action, you have the option to edit the permissions for this action.

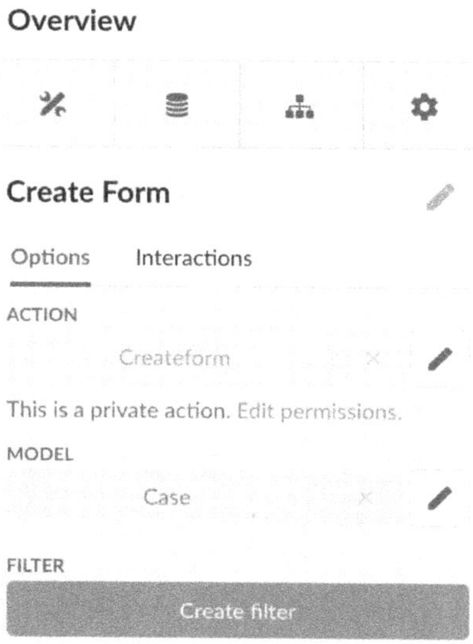

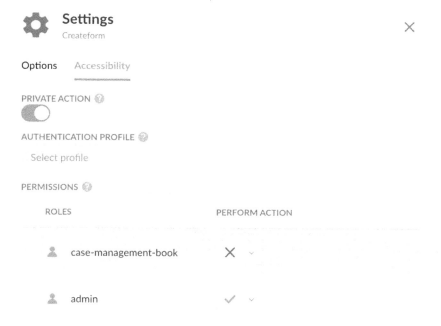

Figure 11.9 – The action linked to the form with the possibility to edit the permissions

It's important to set these permissions correctly, as you only want specific users to be able to perform specific actions. So, let's have a look at the current settings for this action.

Figure 11.10 – The permission settings for an action

First of all, you have the option to make this a public action by simply adjusting the slider of the **PRIVATE ACTION** option. If you make this a public action, every user (including unauthorized users) will be able to execute this action. As we only want a logged-in web user to be able to create a form, we can select a specific authentication profile for this. We also only want web users with the admin role to be able to create a form. Therefore, we will only make this action active for the admin role. If it doesn't work after making these changes, please make sure to check you have the admin role connected to you as a web user.

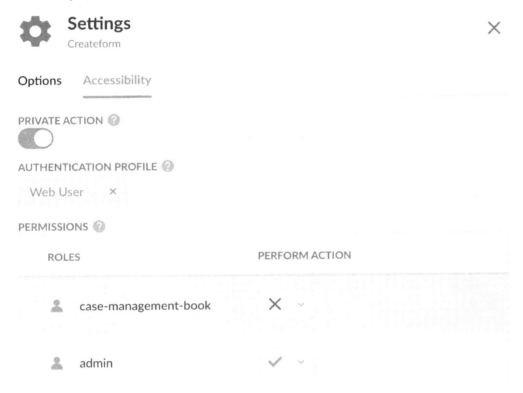

Figure 11.11 – The permissions that need to be set for our action

In the **Options** tab, we can also change the name for our action, as it is now a standardized name. Let's change the name for this action to Create new case.

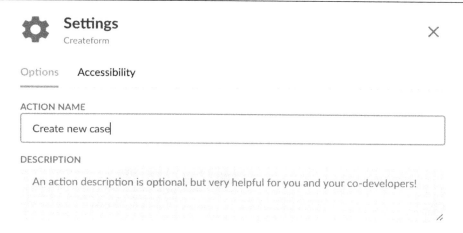

Figure 11.12 – Changing the name of the linked action

After changing this for **Create form**, you will also need to do the same for **Update form** and the Delete form.

The Delete form can be found in **CRUD with slide-out panel**. You can find it by unlocking this wrapper in the component tree from the **Options** menu. After this, simply follow the previous steps to set the permissions for this form as well, and lock it when finished.

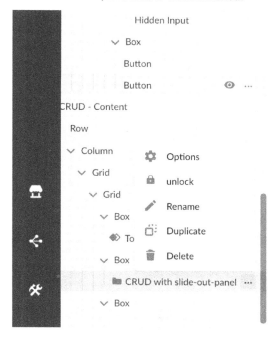

Figure 11.13 – The option to unlock or lock a wrapper in your component tree

When all these steps are done, please make sure you've linked the admin role to your web user, compile the page, and test all the different steps before you continue with the next part.

Actions

We've already created actions before, but they had all been rendered for us. We haven't taken a deeper dive into them yet. This is the time to do so and start understanding actions better as we'll start modifying and creating our actions now.

Actions are basically the backend logic of your application, so where normally a programmer would write backend code to support their application, you can do this with the Betty Blocks actions. Actions are made to automate processes and provide logic in your application. The logic is executed when certain events are triggered: a user clicks a button or sets the timer for some event to happen at an exact time. You can create your own workflow in the action builder by adding action steps. Action steps will serve as a logical construction, as the events will be carried out on conditions that you set yourself.

All the basic steps you would like to do with actions are there by default, such as creating a new record in your model or updating a record. Also, there is a step for conditions, which allows you to make a decision based on data from your models and then to take another flow in your action with a different outcome. All of this is presented and created in a visual flow. See *Figure 11.14*.

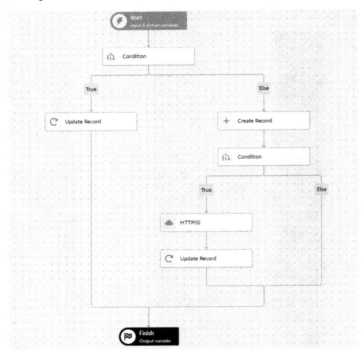

Figure 11.14 – An action with steps

You don't have to write any code here since it's no-code system, but just like with the page builder, there is the option to extend the functionality by creating new functionality with code so that you as a citizen developer or no-coder can do more with actions. You will find out how to do that in the next chapter.

In addition to that, there are also new steps available in the Block Store. There, you can download new steps and add them to your application. All steps are free, so feel free to explore them. You can also use the store to upload your own steps and use them throughout your organization. So, when someone creates a new step, everyone within your organization can use it. There are a lot of advantages to that.

As you can see a step has a start step and an end step. It's important to understand that from both points you can do something with data. At the start step, you can use data that comes from your page (or usually the form that you are submitting). But you can also create your own properties there to access specific data from your data model that you might need later on in your action as well. The same goes for your end step; it can send data back to your page so you can use that data on your page. You wouldn't use this to return whole arrays of data to use in a data table or something, since you can access that data from your data model as well. But you can return, for example, a message about the outcome of your action, so the user can be informed about the success or failure of your action.

All the standard action steps at the time of writing this book are as follows:

- Authentication - Authenticate User
- CRUD - Update, Delete, Create Record
- Debugging - Log Message
- External - HTTP(S)
- Flow - Loop, Condition
- Miscellaneous - Upload File

This part of the platform has recently been completely rewritten, so that's why you might see more options available by the time you are reading this. The previous version of the actions basically worked the same, but couldn't be extended like in this version. Also, this version is much more performant than the previous version. So, it should improve your applications massively, with more steps to come in the near future.

Actions also use permissions, allowing you to ensure that not everyone can run any action. This could be very helpful, for example, in preventing a user from deleting a record when they shouldn't have access to that action. These options can be found in the settings menu at the top of your actions. There, you can find settings that allow you to select which roles have permission to run this specific action. You can also turn off private mode for an action. This means that anyone can click the button that triggers this action. But be careful with doing this as it can be risky.

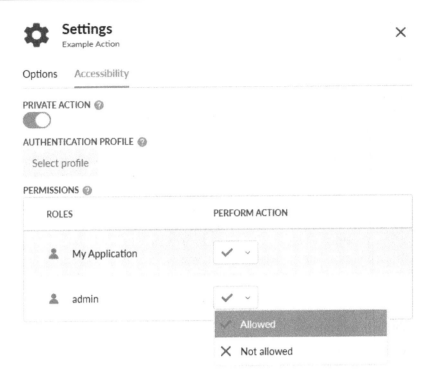

Figure 11.15 – Action permissions

The best way to understand actions is to start using and modifying them, which is what we will do next.

Adding files to the current case

Now that we've covered the basics of creating, updating, and deleting cases, we also want to be able to add important files to our cases. Let's take a look at how we can implement this in our application.

First, we need to select the **Update** tab. To do this, we first need to find the wrapper called **CRUD with slide-out panel** in our component tree. Open the options of this wrapper by double-clicking on it in the component tree, or by selecting **Options** from the menu for the component.

Now that we've opened the options, we need to select **Record view** and the **Update** tab.

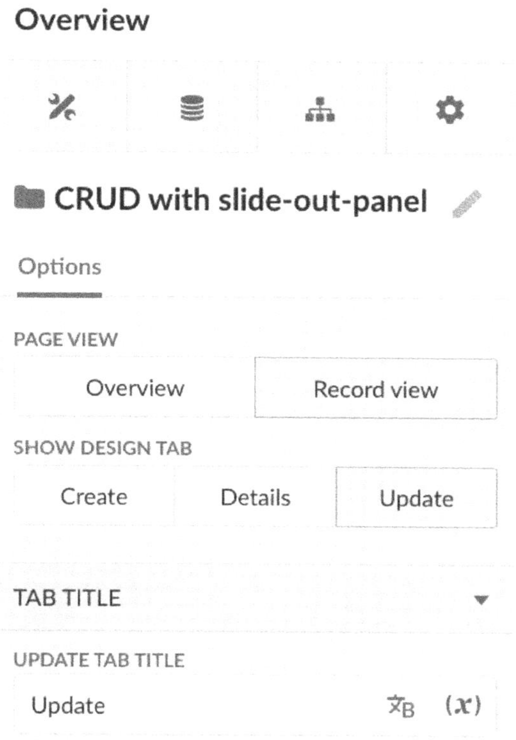

Figure 11.16 – The wrapper options where you can switch between tabs and the general overview

In the **Update** tab, you will see a hidden input. This hidden input is used for the ID of this specific case record. Using this ID, we can make sure the data will only be updated for that specific case.

We will add a button just below this **HIDDEN INPUT**. For now, let's call it `add file`. If you want, you can also add an icon to make it more visual for the end user.

After this, we'll add a dialog as well, by dragging it just below the button. We'll rename the dialog `AddFileDialog`.

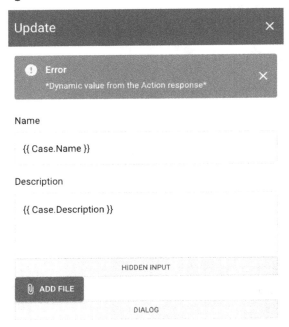

Figure 11.17 – The update form with the ADD FILE button and the new dialog

Now, let's add an interaction to the **ADD FILE** button, as we did before, so we can show the dialog. After this, we'll make the dialog visible by changing the visibility option. For this dialog, we need to make the following changes:

1. Add a **Create** form. In the wizard, link the **CaseFile** model and add the filename and file property to the list before saving.

2. In our **Create** form, a new **Submit** button has been created. Now, let's move it next to the current **Submit** button and, after this, we can simply remove the old **Submit** button.

3. To make sure our **Submit** button works, please drag the box that contains both buttons, within the **Create** form.

4. Change **Title text** to `Add new file` and remove the text just below the title.

5. The last thing we need to do is add a hidden input to the form. We need to link this hidden input to the ID of the current case:

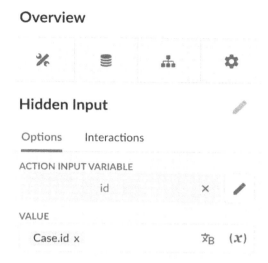

Figure 11.18 – The configuration of our hidden input in the Add File dialog

Now that we've made all the changes to our **AddFileDialog**, it should look like this:

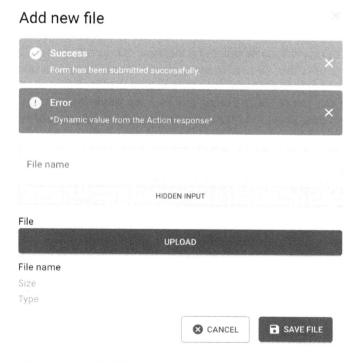

Figure 11.19 – AddFileDialog containing the configured items

Now that we've created the dialog, we need to make some changes to the action in order to be sure the file will be linked to the current case. To do this, we first need to open the options for the **Create Form** and edit the linked action by clicking on the pencil.

Figure 11.20 – Edit the linked action for Create Form

First, we need to select the **Create** record step and select the **Variables** tab. Here, we can add a new variable. We'll select the record variable and give it the name current_case. We also need to link this to the Case model. After this, we need to make sure we select our current case. To be able to do this, we can add filter rules. Let's add a new rule and compare the ID of the record with the ID given in the hidden input. Once you've finished doing this, click the **Apply** button to apply this rule to the new variable. Now, let's click **Save** to add the new variable to our **Create record** event.

Figure 11.21 – The new current case variable containing the new filter rule

Now, let's open the **Options** tab and add a new property to our event so we can assign the **current_case** variable to our case. It's simple; click **Select variable** to select the variable from the list. Then, click **Save**.

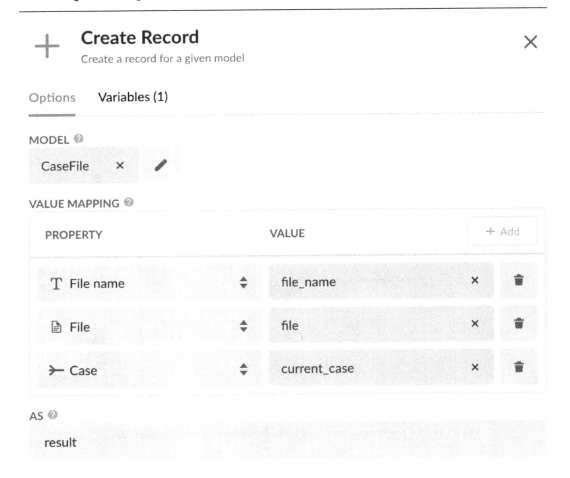

Figure 11.22 – The setup of our Create CaseFile event

Now, simply change back the visibility of your dialog, and you're done. The next steps are adding our notes to a case and showing the files and notes on the details screen.

Adding notes to the current case

We also want to be able to add important notes to our case. Let's take a look at how we can implement this in our application.

First, we need to select our update tab as we did before. To do this, we first need to find the wrapper called **CRUD with slide-out panel** in our component tree. Open the options of this wrapper by double-clicking on the wrapper in the component tree, or by selecting **Options** from the menu for the component.

Now that we've opened the options, we need to select **Record view** and the **Update** tab.

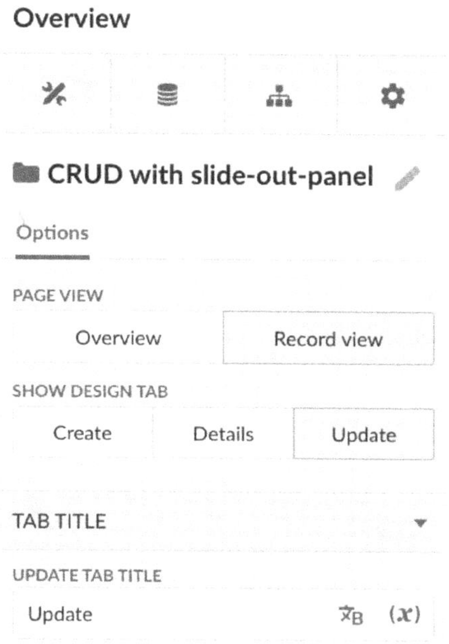

Overview

CRUD with slide-out-panel

Options

PAGE VIEW

Overview | Record view

SHOW DESIGN TAB

Create | Details | Update

TAB TITLE

UPDATE TAB TITLE

Update

Figure 11.23 – The wrapper options where you can switch between tabs and the general overview

We will add a new button just below the existing button and dialog. For now, let's call it Add note. If you want, you can also add an icon to make it more visual for the end user.

After this, we'll add a dialog as well, by dragging it just below the button. We'll rename the dialog `AddNoteDialog`.

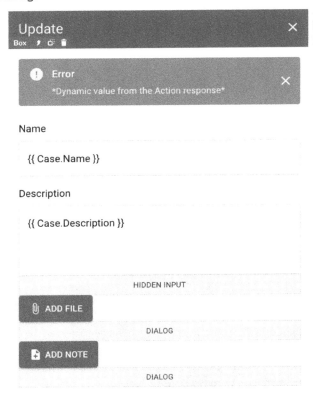

Figure 11.24 – The update form with the ADD NOTE button and the new dialog

Now, let's add an interaction to the **ADD NOTE** button, as we did before, so we can show the dialog. After this, we'll make the dialog visible by changing the visibility option.

For this dialog, we need to make the following changes:

1. Add a **Create form**. In the wizard, link the **CaseNote** model and add the title and description property to the list before saving.

2. In **Create form**, a new **Submit** button has been created. Now, let's move it next to the current **Submit** button and, after this, we can simply remove the old **Submit** button.

3. To make sure our **Submit** button works, please drag the box that contains both buttons, within the **Create form**.

4. Change the title text to `Add new note` and remove the text just below the title.

5. The last thing we need to do is add a hidden input to the form. We need to link this hidden input to the ID of the current case:

Overview
Hidden Input
Options
Interactions
ACTION INPUT VARIABLE
id
VALUE
Case.id x
(x)

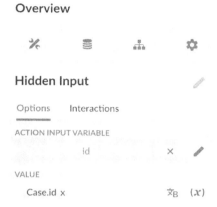

Figure 11.25 – The configuration of our hidden input in the Add Note dialog

Now that we've made all the changes to our **AddNoteDialog**, it should look like this:

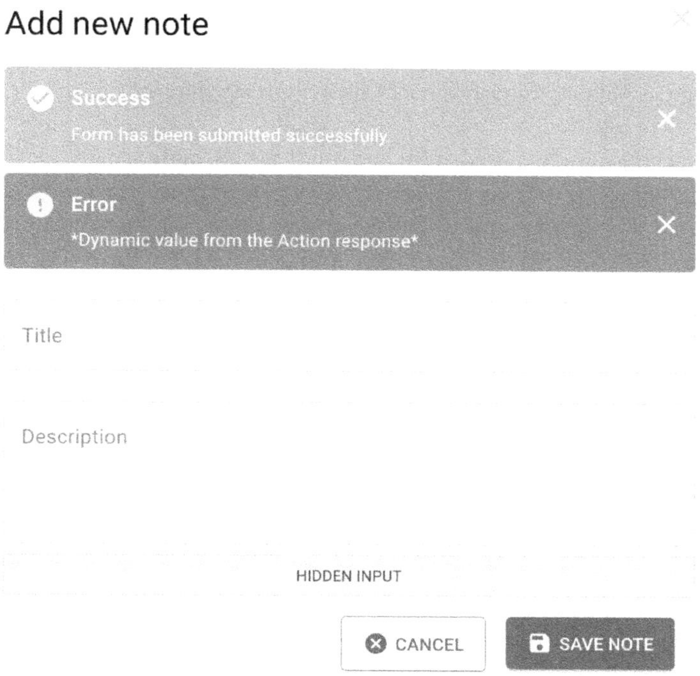

Add new note
Success
Form has been submitted successfully
Error
Dynamic value from the Action response
Title
Description
HIDDEN INPUT
CANCEL
SAVE NOTE

Figure 11.26 – AddNoteDialog containing the configured items

Now that we've created the dialog, we need to make some changes to the action in order to be sure the note will be linked to the current case.

To do this, we first need to open the options for the **Create form** and edit the linked action by clicking on the pencil.

Figure 11.27 – Edit the linked action for Create form

First, we need to select the **Create record** event and select the **Variables** tab. Here, we can add a new variable. We'll select the record variable and give it the name `current_case`. We also need to link this to the Case model. After this, we need to make sure we select our current case. To be able to do this, we can add filter rules. Let's add a new rule and compare the ID of the record with the ID given in the hidden input. Once you've finished doing this, click the **Apply** button to apply this rule to the new variable. Now, let's click **Save** to add the new variable to our **Create record** event.

Figure 11.28 – The new current case variable containing the new filter rule

Now, let's open the **Options** tab and add a new property to our event so we can assign the **current_case** variable to **CaseNote**. It's simple; click **Select variable** to select the variable from the list. Now, click **Save**.

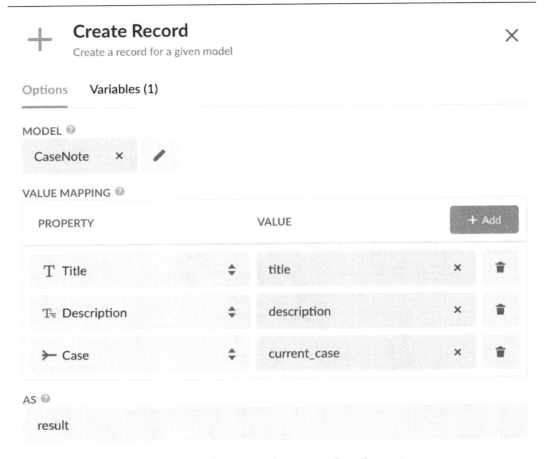

Figure 11.29 – The setup of our Create CaseFile event

Now, simply change back the visibility of your dialog, and you're done. The next steps are showing the files and notes on the details screen.

Showing the case details page

Now that we've added the option to add files and notes to our case, we also want to show these items when we open the case details.

First, we need to select our **Details** tab. To do this, we first need to find the wrapper called **CRUD with slide-out panel** in our component tree. Open the options of this wrapper by double-clicking on the wrapper in the component tree, or by selecting **Options** from the menu for the component.

Now that we've opened the options, we need to select **Record view** and the **Details** tab.

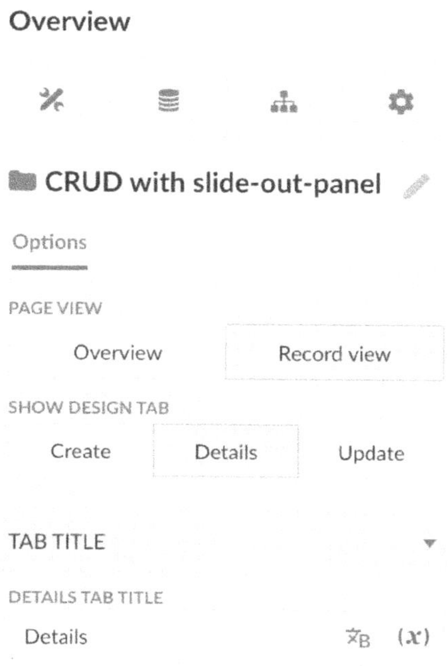

Figure 11.30 – Selecting the Details tab

As you can see, the tab is currently not that wide, and therefore it can't contain all the information we want to show. So, let's make it wider first. In order to do this, we need to select the **Drawer** component in our component tree. This should be the first item on the list. Let's open the options for this component. Now, we can make the drawer wider by setting the drawer width. Let's change this to 800 pixels, or wider if this suits you better.

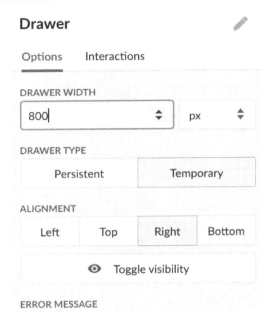

Figure 11.31 – Changing the width settings for our drawer component

Now that we've changed the width, we will add an expansion panel to our details screen. Make sure it is added to the data container. After adding this, give it the title `Overview files`. By default, we'll show this expansion panel expanded.

Now, we'll add a datalist to this expansion list. This datalist has pagination included by default, and we can also change the number of items we want to display. We need to link this datalist to our **CaseFile** model. As our expansion is already dragged into the data container, we can now simply create a filter for the datalist where we will show only the files linked to the current case.

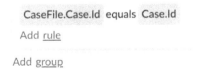

Figure 11.32 – The filter used for our datalist

After this, drag a text component onto the datalist. Select the text component and link the content to the case file. To make sure we can download the file, set the text component to link to an external page and set the URL to the file property in our **CaseFile** model.

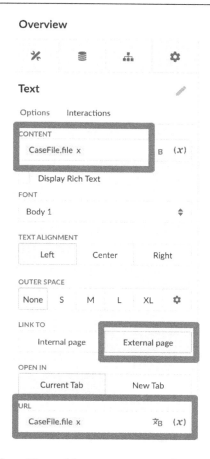

Figure 11.33 – The settings of the text component for our overview of files

Now, we can also add a notes overview to the application. Again, we will add an expansion panel to our details screen. Make sure this is also added to the data container. After adding this, give it the title `Overview notes`. By default, we'll show this expansion panel expanded.

Now, we'll also add a datalist to this expansion list. We need to link this datalist to our **CaseNote** model. As our expansion is already dragged into the datacontainer, we can now simply create a filter for the datalist where we will show only the notes linked to the current case.

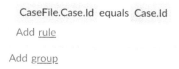

Figure 11.34 – The filter used for our datalist

After this, drag a text component onto the datalist. Select the text component and link the content to the title of our note.

Figure 11.35 – The settings of the text component for our notes overview

The end result of your details page should look like this:

Figure 11.36 – Our final details screen

Now that we've completed our details page, simply compile the page and see whether everything works as expected. If something isn't working or you are unable to select the ID of your case in the filter, double-check to see whether the components are all within your datacontainer.

Changing the status

Now that we've completed all the basic functionality, we also need to add the ability to change the status of a case.

First, we need to select our **Update** tab. To do this, we need to find the wrapper called **CRUD with slide-out panel** in our component tree. Open the options of this wrapper by double-clicking on the wrapper in the component tree, or by selecting **Options** from the menu for the component.

Now that we've opened the options, we need to select **Record view** and the **Update** tab.

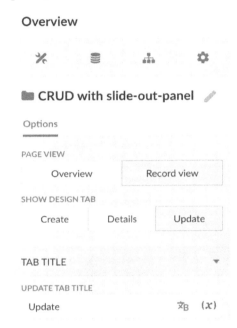

Figure 11.37 – The wrapper options where you can switch between tabs and the general overview

To create a button with a linked action, we can search for the action button and drag this on our form. After dragging it on the form, select the action button and open the action in the settings for this button.

We need to pass the ID of the case to our action. To do this, select the ID in the **PASS VALUE TO ACTION** field. Then, open the action by clicking on the pencil icon.

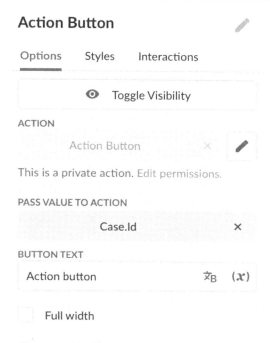

Figure 11.38 – The settings of our action button

Within the action, we need to add an **Update record** event. In this **Update record** event, we first need to add the **current_case** variable, as we did before in our other events. We'll select the record variable and give it the name `current_case`. We also need to link this to the Case model. After this, we need to make sure we select our current case. To be able to do this, we can add filter rules. Let's add a new rule and compare the ID of the record with the ID given in the hidden input. Once you've done this, click the **Apply** button to apply this rule to the new variable. Now, let's click **Save** to add the new variable to our **Create record** event.

Figure 11.39 – The new current case variable containing the new filter rule

Now, open the **Options** tab of the **Update record** event and select the **current_case** variable as your record. After this, select the **Status** property from the dropdown and select the status you want to set it to, for example, **Closed**.

Figure 11.40 – Our update event for changing the status

Tracking the history of your case

Now that we've added our status change, we can also add the ability to track historical changes for our case. You can follow these steps for every part of history you would like to track in your application. For now, we will only track the history of the status change.

When adding this functionality to our action, we have to make sure that the create event that we will add is added just before the update event itself.

In this create event, we have to retrieve our current case using the variables section.

Summary

In this chapter, we finished our case management application. We added authentication to our case management application so that we can log in with a user.

We used the CRUD with slide-out template as a basis for our case management application, so we don't have to create the whole CRUD flow ourselves, as most of it is already predefined for us. Also, we learned how to unlock and use a wrapper to get around faster in our page template.

We took our first steps into actions in this chapter. We learned how to change the permissions of an action so we can control who launches it. We also went through the basics of actions: what they can do and how we can extend them. We added some extra functionality to our actions, so we can add notes and files to our case, track the history of it, and change our status. Most of this required adding some extra functionality to our action, so you learned how to modify those.

In the next chapter, we'll go over the pro-coder features of the platform. You will learn how to extend the component set in the page builder and add new steps to your actions. The following chapter is focused on pro-coders, so they can assist no-coders and citizen developers with extra functionality that they might need when creating applications.

Questions

1.   If you need to make a choice in an action, which step would you use and why?

2.   What is a wrapper used for in a page template?

3.   With which step in an action can you create a new record in your model?

Answers

1.   The conditional step. This step will split up your flow into two flows, so you can do different things in both flows.

2.   A wrapper is a part of your page template that makes it easier to navigate through it. You don't have to manually find that specific component in the component tree to alter it, but you have access to all the main components from your wrapper.

3.   The Create record step. It allows you to select a model and then assign the properties you want.

Part 4:
The Pro-Coder

In this part, we'll demonstrate some pro-coder options in the platform so that citizen developers can understand what pro-coders can do for them, but also so pro-coders understand what they can do in the platform and how to do it. This part has the following chapters:

- *Chapter 12, The Pro-Coder Features*
- *Chapter 13, The Back Office*

12

Pro-Coder Features

In the first two chapters, we talked about the pro-coder. In this chapter, we'll dive deeper into what a pro-coder can actually do in the platform with some examples. The two main pro-coder features of the platform that we'll discuss are the component set and the action functions.

The main topics of this chapter are as follows:

- The component set

- Structure of a component and prefab

- Using a new component

- Action steps

By the end of this chapter, you'll have a better understanding of what the pro-coder features are and what you can use them for. This guide will not go into detail about how to create them (for pro-coders, this should not be a hard task), since I will assume you already know how to code. The focus is on understanding the way Betty Blocks implements these features, so that you, as a pro-coder, can utilize them to their full potential.

The main language for using the pro-coder features is JavaScript/TypeScript and React. The component set uses the React framework, while the actions use JavaScript. Having knowledge of GraphQL could help as well since the data and the metadata that the Betty Blocks platform provides come through GraphQL. Let's start with diving deeper into the component set first and how you can extend this.

The component set

So what is the component set exactly? This is the default set that Betty Blocks provides with all the components that are in the **Page Builder** by default. So, things such as columns, forms, input, and buttons can be found there. But with the code of the component set itself, you can modify these components, so they can do more than the standard components that Betty Blocks offers. The standard set is already pretty complete, but there might always be an edge case that you would like to make work. In most no-code platforms, you are limited by the standard set, but here, you have basically all the freedom that the web offers.

As mentioned in the introduction, the component set is built in React. This means you need to have a basic understanding of how React works. It uses React 16.8+ functional components, so having an understanding of React version 16.8+ is also recommended along with TypeScript.

You can find the component set on GitHub. Go to `https://github.com/bettyblocks/material-ui-component-set` to download the set. This is only the first part because to run the component set in the platform, you'll need the Betty Blocks **command-line interface** (**CLI**). This is also available from GitHub (`https://github.com/bettyblocks/cli`), and with this CLI, you can build the component set and connect it to your Betty Blocks application, so the changes you've made become available in your page builder as well. Learning how to connect it to Betty Blocks will be explained in the *Using a new component* section later in this chapter. But first, we need to install the CLI to get the component set up and running.

The Betty Blocks CLI

The Betty Blocks CLI is currently used for two different things – first, the component set, and second, the action functions. So, once you've installed the CLI, you are good to go for developing both. There is a specific option when installing the CLI to only install the part for the component set if you have no desire to develop for the functions, as for the functions, you'll also need to install the following:

- Make
- G++
- Python

For the component set, these are not needed. This has to do with the fact that the action functions run on an isolated VM. More information about isolated VMs can be found in the Wiki on GitHub for the CLI.

Another important piece of software that needs to be installed is a recent version of Node.js. Since this changes over time, please check the Wiki for the exact version that you need at this point.

So what does the CLI do for the component set? Once you start changing or building components, they'll need to be built and prepared so that Betty Blocks can read them. This is what the CLI can do for you with specific commands. Actually, the only thing you need to do is run the CLI on a component set that you have downloaded, and every time you make a change, it will automatically build a new version, so there is not a lot of knowledge needed there. You only need to get your CLI installed and the component set downloaded. You can install the CLI now using the steps on the GitHub page for the CLI (*Figure 12.1*):

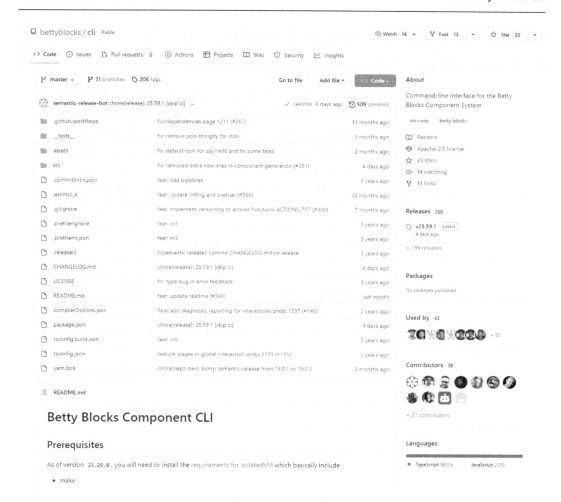

Figure 12.1 – The Betty Blocks CLI GitHub page

I'd recommend installing the whole CLI if you want to use the actions functions as well. After you've installed the CLI, you can either clone the component set from GitHub in your favorite text editor (for example, Visual Studio Code) and then in your CLI in the component set folder, run `yarn install` to install all the dependencies for the component set. Once that has been installed, you can type `yarn run dev`, which should start the development server. Now, if you make changes to the component set, it will automatically rebuild it. Now you are all set for making your first changes to the component set.

Structure of a component and prefab

So before you can start building a new component, you need to understand how they work. Since the components need to be usable in the platform, it's not only the component itself you need to create; you also need to define what component options the no-coder and the citizen developer have available to them, so they can drag and drop the component on to their canvas in the page builder and then use the component options to set it up to their liking or make changes to the style of the component. So we need to take care of that as well in the component. This is why a component consists of two different things: the component and the prefab.

Using a new component

So let's use a component in the Betty Blocks platform itself now. We're not going to build a very complicated component, but just a simple example to show how it works. Make sure you have the CLI installed. If you don't have it installed yet, go back to the first section of this chapter to see how to install it.

To make things easier, clone the Material UI component set from the Betty Blocks GitHub page. Let's use `yarn install` to get it ready. You can do this from your terminal or the terminal in your favorite editor. Once it's done, get it up and running by running `yarn run dev` (*Figure 12.2*):

```
[nodemon] 2.0.20
[nodemon] watching path(s): src\**\*
[nodemon] watching extensions: js,ts,tsx
[nodemon] starting `yarn build && yarn start`
$ bb components build
@betty-blocks/cli update available from 25.56.0 to 25.59.1
Success, the component set has been built
$ bb components serve -p 5002
@betty-blocks/cli update available from 25.56.0 to 25.59.1
Serving the component set at http://localhost:5002
```

Figure 12.2 – Your Betty Blocks CLI up and running in the terminal

Let's start by adding a new file to our component's folder. You can create it in `src/components`. Let's name it `exampleComponent.js`. Put the following code in this file:

```
(() => ({
  name: 'exampleComponent',
  type: 'CONTENT_COMPONENT',
  allowedTypes: [],
  orientation: 'HORIZONTAL',
  jsx: (() => {
```

```
      const { textContent,
  fontType,
  textAlignment,
          } = options;
      const { useText } = B;

      const parsedContent = useText(textContent);

      return <div className={classes.content}>{parsedContent}</
  div>;
    }) (),
    styles: () => () => {
      return {
        content: {
          display: 'block',
          padding: 0,
          whiteSpace: 'pre-wrap',
        },
      };
    },
  })) ();
```

You might see an error appearing in your console right now; ignore it. It will fix itself later.

This is a simple text component, it will show some text that you can define in the content option of the component and show it on your page. Out of the box, this will be both content from your data model and static text that you can type yourself.

Let's go over the code. In its basis, it is React code, but with some Betty Blocks helpers, so we'll focus on the parts that are specific for Betty Blocks components.

First, we have name, which will connect the component to the prefab, so it needs to be unique and identifiable. Next up, we have type, which determines the type of your component. You can set the allowedTypes option (which comes next) with the specific type, which will allow it to be nested. If you keep allowedTypes empty, all components will be allowed inside this component.

Next, we have the orientation, which has to do with the default orientation of the component, so it scales well in your canvas. You can set the orientation to either horizontal or vertical.

The biggest and most important one is the JSX. Here comes the code for your component. Within the JSX, you can see two helpers already there, which are important as well: the options and the B (Betty) helpers.

The options refer to the options that have been set in the prefab. Since we haven't created a prefab yet, it might look a little strange, but it will make sense once we create the prefab. You can have 1 option, but usually, you have more than 1 – up to 50, for example. The option here is called `content` and will hold the content for our example simple text component.

The B helpers are some built-in functions that make your life easier while building components. In the example you see right now, `useText` is a helper for parsing the text correctly from `options.content`, without having to write a function yourself. The most important functions here are those for retrieving data from your models and those that can help you transform data, such as `useText`. All of these functions can be found on the GitHub page of the CLI (`https://github.com/bettyblocks/cli/wiki/Components:Component-Helpers`).

Last, we have the styles, which is the part that allows you to add custom styling to your component. Also, these styles interact with the options in your prefab and with the theming that you can set in Betty Blocks. You can find examples of this on the GitHub page of the CLI.

These are all the basics you need to understand to create a component. Of course, you can use React hooks here as well and also create your own interactions for your components.

The prefab

The prefab is a file where the structure, options, name, and icon of a component are defined. It can refer to one or many components, so a prefab is not always just one component. For example, the `box` component is basically a `div` in HTML, which is only one component in a prefab as well. But `dataTable` consists of more than one component: it has the data table itself and also the `dataTable` column defined in its prefab.

A prefab determines the options your no-coder or citizen developer can set on your component. It also can hold more than one component, so you can create more complex components with prefabs by combining components. It is most important to understand the basics of a prefab, so you can create these. We are not going to create a complex prefab here; we're just going to use a few options that we've basically already defined in our component in the previous section.

In a prefab, you usually don't define a lot of pure JavaScript code; most of the prefab is JSON. The only JavaScript code that you might add to a prefab might be related to what is called `before create`. This is an option that allows you to give the user an interface to set up a series of screens with a stepper that guides the user to set up their component. For example, in `before create`, you can select a model and assign the properties before the component is created. The data table is a good example of this. When you drag it onto the canvas, you'll notice that it will have some dialogs popping up first to help you set up your data table. This way, the user doesn't have to drag a lot of columns onto the canvas, for example. Most page templates also use `before create`, so that users can do the setup of their page before it's created for them. Have a look at these files in your component set to further understand this concept. We won't use it in our example here, but it's something that can be very useful if your component is going to be used a lot. You can reduce the time for the user to set up the component by adding `before create`.

What `before create` does is introduce code to show a dialog to the user and use the outcome of that dialog to set up the options that you have defined in the prefab. Instead of adding default values to your options here, these options will be variables that have been defined in `before create`. We'll have a look at how you define these options soon, but since you'll encounter the `before create` dialogs a lot in your components and page templates, I thought it would be useful to understand where those come from before we dive into the other parts of the prefab.

Let's add a new file called `exampleComponent.tsx` to our `src/prefabs` folder.

Next, you can see the code for the prefab; copy this into the file we just created:

```tsx
import * as React from 'react';
import {
  component,
  prefab,
  variable,
  Icon,
  font,
  buttongroup,
} from '@betty-blocks/component-sdk';

const attributes = {
  category: 'CONTENT',
  icon: Icon.TextInputIcon,
};

const options = {
  textContent: variable('Content', { value: ['This is my first
component!'] }),
  fontType: font('Type', { value: 'Title1' }),
  textAlignment: buttongroup(
    'Align',
    [
      ['Left', 'left'],
      ['Center', 'center'],
```

```
    ['Right', 'right'],
  ],
  { value: 'left' },
 ),
};


export default prefab('ExampleComponent', attributes,
undefined, [
  component('exampleComponent', { options }, []),
]);
```

Let's go through this code piece by piece to understand it better. First, we see the import of React; this is needed since it's React with TypeScript. Then in the next part, we see another import that imports some helpers from our own component set SDK. More information about the SDK can be found on the Betty Blocks GitHub page: https://github.com/bettyblocks/cli/wiki/Components:Home.

Next, we have the attributes for the prefab; these attributes are for configuring the category we want the prefab to show up in, in the component set overview in the page builder. The icon refers to an icon that is available within this component set. All the icons can be found here, under icons: https://github.com/bettyblocks/cli/wiki/Components:Types.

As mentioned in the *The component set* section of this chapter, here are the options again. As you'll notice, these are the same options as in the component and are defined here. Let's take a look at them.

First, we have the textContent option. Since content can also be a variable from your data model, it's defined here as a variable, but you can set the value to text here as well, as is done in this example. As you might notice, the variable is a helper from the SDK.

Next, we have the font type. This just defines which font should be used by default. And under that, we have the align option. This one should be buttongroup, which is taken from the SDK where it has been defined, so you don't have to worry about that; you just have to set its name, labels, and values. And last, you set the default value again in value.

As you can see, it's quite easy to set up the options for a prefab. A prefab can have from no options to as many options as you need. You can even use categories to organize your options better so your users are not overwhelmed.

The last part is the export part, which exports the prefab, gives it a name, and sends the attributes with it that we defined previously. undefined here is the place where you would normally add before create, but since we haven't defined that here, it's undefined. And then next, you define which component it relates to and pass on the options to that component. A component can have descendants, or child components, which you can define in the array. If you want to see an example of that, I can recommend looking at the stepper or tabs prefab, for example.

Now that we have defined the prefab, we can load our new component into our component set in the Betty Blocks platform.

Running your local component set

To run your local component set in the platform, you'll only need to set up one of two settings. Why two? Because there are two ways of doing it. There is one way that allows you to set it globally for all pages or another way in which you can load it for a specific page. We'll go over both options, as the global one, in particular, is not visible as an option in the platform.

In the **Builder** menu, go to **Tools** (represented by the hammer and wrench at the bottom); a submenu will open, in which you will see **Configurations**:

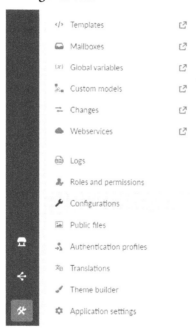

Figure 12.3 – Configurations in the Tools menu

Configurations is where you set up configurations for actions, but also for setting up pro-coder-specific features that are not visible to the citizen developer and no-coder. Here, click on **New configuration** and you'll get a screen where you can enter a name and a value. The name should be `component_set` and the value should be set to the host that your CLI is serving the component set at – you can check that in your terminal. By default, this is `http://localhost:5002`.

If you go to your page builder now and search for the example component, you should see it appear in your component set on the left side of the screen. The whole component set is now loaded manually. You can disable it again, by removing the configuration.

The other option is more straightforward, but it only works on a specific page, so you'll need to set it manually on each page. Make sure you remove the configuration that we've just added and go to any random page in Betty Blocks. Go to the **Settings** tab on the left side of your screen. There, you should see an option called **COMPONENT SET**, which should be set to default. This means that it will load the default component set provided by Betty Blocks from their server (see *Figure 12.4*):

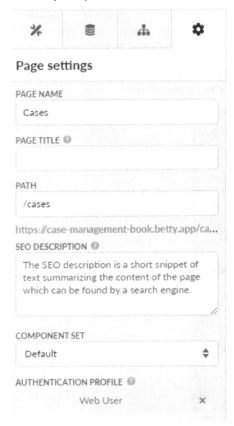

Figure 12.4 – The default component set in the Settings tab

If you expand the dropdown, you should see an option called **Local**. This should use the `http://localhost:5002` address. If you are using any other address because of some other port, you need to use the **Custom** option. With this option, you can type the address manually. Now, if you click **Save**, it should load the component set on this page from your localhost as well. That's all about how you can use custom components in your apps. Let's have a look at the action steps now.

Action steps

One of the newest features of the Betty Blocks platform is that you can add your own action steps to the platform. This basically means that you can create backend logic with code for your no-coders and citizen developers. The platform already features out-of-the-box steps for your actions, but if you need something that isn't supported, you can create it yourself. Does that mean you can do anything? The short answer is, no. But you can do a lot, and in the future, you'll be able to do more and more here. As mentioned before, the action steps are based on Node.js, but of course, you don't have access to the filesystem, so it's more limited than normal Node.js because of security.

We'll go over the basics of the action steps here, but we aren't going to create our own action step as with the components, because you can do so many different things here compared to components that it wouldn't make sense. Also, at the of writing this book, it is still in beta, so there might be some changes as well, so be aware of that.

Just like with the component set, there is a GitHub repository with all the current action steps available in code. You can find that here: `https://github.com/bettyblocks/block-store-action-functions`.

The CLI for the action steps is exactly the same one you use for the component set; it's just a different set of commands you can use. So there is no need to install the CLI again. There are a few extra requirements for actions, which you'll need to install, as mentioned earlier in this chapter. You'll need to install the following additions:

- Make
- G++
- Python

Without these, the actions part of the CLI won't run.

So what can you do with action steps? You can call a specific web service, for example, that needs some specific functionality that is not available out of the box. Also, steps can have an output variable, so any step that you create can output something that can be used in the next step or even a few steps ahead. So when you are creating a step that calls that web service, you can use it in multiple steps in your action.

Another use case might be that you need to modify some of your data in a specific way that is not available in actions at this moment; you can also create this yourself now. You can take data from your data model, change that in your action step, and then output it again. Here is an example of such a step:

```
const capitalized = (value) =>
  value.charAt(0).toUpperCase() + value.toLowerCase().slice(1);
```

```
const parameterized = (value) =>
  value
    .trim()
    .toLowerCase()
    .replace(/\s/g, '-')
    .replace(/[^a-zA-Z0-9 -]/g, '');

const textTransform = async ({ value, transformation }) => {
  let result;

  switch (transformation) {
    case 'downcase':
      result = value.toLowerCase();
      break;
    case 'upcase':
      result = value.toUpperCase();
      break;
    case 'capitalize':
      result = capitalized(value);
      break;
    case 'parameterize':
      result = parameterized(value);
      break;
    default:
      result = value;
  }

  return { result };
};

export default textTransform;
```

This action step can transform some text and lowercase characters or uppercase characters, capitalize them, or parameterize them. So you can add more options to this step, for example. As you can see from the code, it's just a simple JavaScript function. It returns a result, which you can use in your action in other steps.

If you go to this step on GitHub (`https://github.com/bettyblocks/block-store-action-functions/tree/main/functions/text-transform/1.0`), you'll see that there is also a `function.js` file. This is almost like the prefab from the component set; here, you define all the options for the step for the UI in the actions. All the definitions to be able to build this can be found here on GitHub: `https://github.com/bettyblocks/cli/wiki/Functions:Definitions`. This should allow you to build up the UI that you'll need for your action step.

There is one big difference between the component set and the action steps and that is how you deploy your action steps to the platform. With action steps, this is done through the CLI; there is a `publish` function available (`bb functions publish`). This function will ask you to which application you want to upload it and you'll have to provide your login credentials for the platform. This way, it will upload the action step directly to the platform. It should become available directly in your sidebar in the actions when it's completed, so this makes that process quite easy.

With this section, I hope you'll be able to start building your own action steps now. On GitHub, there is a lot of information available on the action steps and a lot will be added in the future as well. Before you start building your own, check out the **Block Store** first, if it's not already available; this should make your life a lot easier as well.

Summary

In this chapter, we've talked about the pro-coder features for building your own components for the page builder and for the actions. This should allow the pro-coder to add functionality to the platform that might not be available out of the box. These are usually specific functionalities that are not very common. The CLI is available on GitHub to enable the pro-coder to do this and to upload the component set or action step to the platform.

Both use JavaScript, while the component set is based on React and the action step is based on Node.js, but with a few limitations.

In the next chapter, we'll go over the back office in the older IDE of Betty Blocks. It's a deprecated feature but still used widely by Betty Blocks users.

Questions

1. What does the CLI do?

2. What is the difference between the component set and the action steps?

3. Where can you find a lot of information about these pro-coder features?

Answers

1. The CLI allows you to use the Betty Blocks libraries to develop your own components or action steps. It also allows you to upload them to the platform.

2. The component set is aimed at a component for your web page, so it's based on frontend functionality. The action steps are focused on the backend, such as writing data to the data model.

3. You can find this on GitHub at `https://github.com/bettyblocks/cli/wiki`.

13
The Back Office

In this chapter, we'll talk about the Back Office of **Betty Blocks**. It's very specific to the Betty Blocks platform, and it has a different URL than all the other parts of the Betty Blocks platform that we've seen before now. What is the Back Office exactly?

The **Back Office** is a simple **create**, **read**, **update**, and **delete** (CRUD) interface that allows you to interact with your data very easily in your data model. You can also view this as the administrator environment. In the previous chapter, we talked about data models but had no way to see the data inside them or add any data to our models.

That's where the Back Office comes in. Of course, the platform also has the page builder, which allows you to do the same things as the Back Office and can be made more customer-facing. But the Back Office does these things faster, albeit with some limitations. By limitations, I mean that the way the Back Office looks is always the same. You have control over which models you want to add and in which order you would like to show your data and your properties, but you can't make any changes to this layout. This allows you to set this up very quickly, and you have full control over the data but much less control over the way your pages look. But, for that, we have the page builder.

We'll go over the following topics in this chapter:

- Adding your first model to the Back Office
- Adding a subview to your view

The topics covered in this chapter will help you understand how to use the Back Office and what it can offer you while building your application. Let's get started.

Adding your first model to the Back Office

We'll see an overview of the Back Office in this section, then we'll start making some changes to the Back Office ourselves by adding a new view and a subview to it.

Exploring the Back Office

In this section, we'll explore the Back Office. I'll show you where to find it and what it is used for primarily.

So, where can you find the Back Office on the platform? In the builder bar, on the left, it's the fourth icon from the top (see *Figure 13.1*):

Figure 13.1 – The Back Office icon in the builder bar

Click on the Back Office icon and it should take you to the Back Office. As you might notice, it opens a second tab in your browser and loads the Back Office in a different environment. The Back Office is one of the first parts that was built for the platform and is still in the platform's Classic interface, which loads from a different URL. But no worries, everything that you've done in the new interface is also available here. Most features are being transferred to the new interface, but the Back Office hasn't been transferred yet. That's why we'll use it in this interface. Your screen should look like *Figure 13.2*:

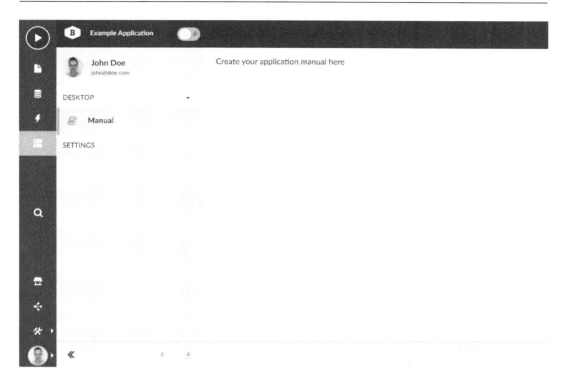

Figure 13.2 – The Back Office

Now that we are in the Back Office, let's have a look around there. When you entered the Back Office, you landed on **Manual**, which by default is empty. We won't use this **Manual** option, but it's something that you could use to inform your users about the setup of your Back Office, for example.

We'll discuss later how to edit **Manual**. If you click on **Settings**, two more options should appear. These *options* here are called **views**. A view represents a model that can show you the data inside the model from your data model. Let's click on **Users**. As you might have noticed in the data model earlier, there were models called **Users** and **Roles**, which are present here too. These are default models that come with any new application and are also present in the Back Office by default. If you click on the **Users** view, you'll see that the right part (the grid) of the screen changes. It should look like *Figure 13.3*:

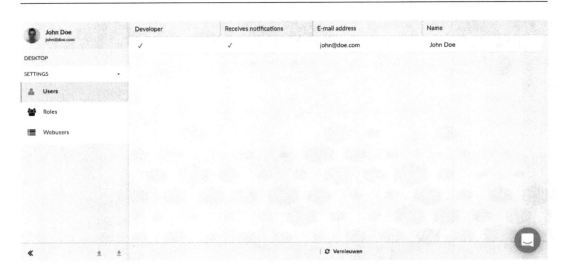

Figure 13.3 – The Users view

The top part of the grid shows you the headers of all the columns of your grid. You'll see there, for example, **Name** and **E-mail address**. This grid shows you all the users of your application. These users are your builder users, so they have access to the part of Betty Blocks where you can build applications. Of course, these users can also be used for your frontend, but we'll get into that in the page builder chapter. Click on the first row of the grid, where it says your name and/or email address. This should open up the details of this record in a sliding pane on the right side of the screen (see *Figure 13.4*).

A record is a single entry in your database. This means that every time you input a new customer in your database, for example, you get a new record.

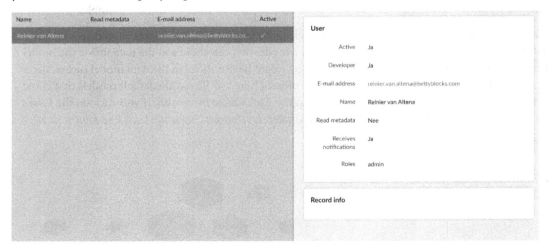

Figure 13.4 – The sliding pane opened after clicking on the record

Here, you can see the details of your record. At the top, you can also see a button for editing these details. The Back Office allows you to easily access all your data like this and make changes to it. So, let's have a look at the most important feature of the Back Office, which is being able to add your own models to the Back Office, so you can add data to your models from the Back Office. Then, we'll also learn step by step how you can use the rest of the Back Office.

Making changes to the Back Office

In order to make changes to the Back Office, we need to turn on **Builder Mode**. This can be done in two ways. You can click on the slider in the top-left corner with the wrench inside the slider, or you can press the E button on your keyboard. Both of these options also turn off Builder Mode. When using the E button, one thing is important: ensure you are not currently in any field where you can type. If you are, this won't trigger Builder Mode. Once you've activated Builder Mode, you should see a yellow bar at the top of your screen saying **BUILDERMODE (PRESS 'E' TO EXIT)**, as shown in *Figure 13.5*:

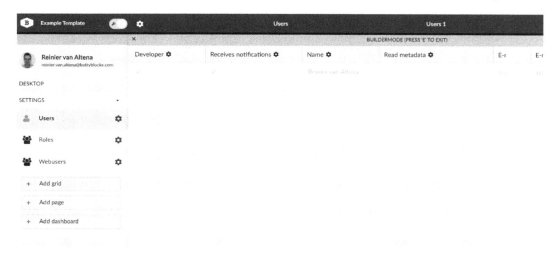

Figure 13.5 – Builder mode activated

In the menu on the left-hand side, you'll see three options:

- **Add grid**
- **Add page**
- **Add dashboard**

We'll only focus on the first one. For the other two options, there are better ways of doing that now, so we'll skip those in this book. We would like to add our new grid to our desktop. So, let's click on **DESKTOP** in our menu, then click on the **Add grid** button (see *Figure 13.6*):

DESKTOP
Manual
+ Add grid

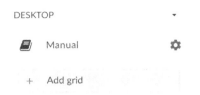

Figure 13.6 – Add grid in the DESKTOP section

Once you've clicked on the **Add grid** button, a dialog should open, as shown in *Figure 13.7*:

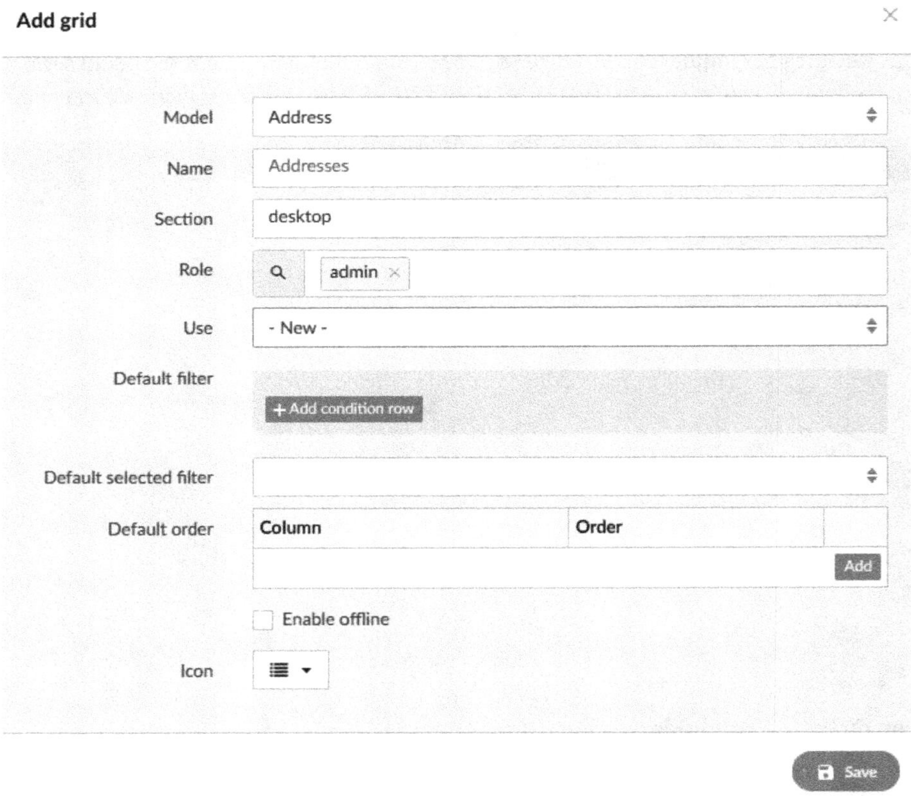

Add grid
Model
Address
Name
Addresses
Section
desktop
Role
admin
Use
- New -
Default filter
+ Add condition row
Default selected filter
Default order
Column
Order
Add
Enable offline
Icon
Save

Figure 13.7 – The Add grid dialog

In this dialog, we can select a model and transform it into a grid or view. If you open up the model selector, you should see all the models that you've created. In this case, we'll start by choosing the **Customer** model. As you can see, the name automatically changes to the model's name as well. If you've used an English name, it should *plurify* the name automatically as well, in our case, to **Customers**.

The **Section** field is the section where the grid/view will appear, in the Back Office (*Figure 13.8*, for example). We'll leave that desktop for now. The **Role** option is used for setting the permissions for this grid/view so that only users with that specific role can access the view. For our example, we'll leave this option set to **admin**. The next field is called **Use** and it's empty by default. The name of the option might be a bit strange, but this is the template for your grid/view. Once you've set up your first grid/view, you can reuse that grid/view again in another one with the same model, but with different permissions. So, we'll leave this option set to **New**.

The filtering options can help you to set some default filter options for this grid/view. Right now, we don't need that, so we'll leave all of those fields empty. We'll talk about filtering at the end of this chapter, so you'll learn all about that then. The ordering options are to change the default ordering of your data in this grid/view. By default, it's ordered on the ID of your records. For this example, we'll leave it as the default.

Lastly, there is **Icon** in the **Add grid** dialog (see *Figure 13.7*). You can use this to set an icon for your grid/view. Choose one that you like for this grid/view. Once you are done, hit **Save** and your first grid/view should be created. Your menu should now look like *Figure 13.8*:

Figure 13.8 – The Customers grid/view has been added

Now, if we look at the grid itself on the right side, we'll see that most of our properties have been added here. You might also have noticed that all the columns have a cog icon beside them that allows you to edit them. You can change the column name here and set them to allow ordering. Not only that, but you can also drag them into your own order. Right now, they are automatically ordered (see *Figure 13.9*):

Figure 13.9 – Initial ordering

Let's drag **Last name** between the **First name** and **Is active** columns. Let's make the order as follows: **First name**, **Last name**, **Date of birth**, **Revenue last year**, and **Is active**. After that, it should look like *Figure 13.10*:

Figure 13.10 – Reordered columns

This reordering allows you to display your data in such a way that it's easy to find a user record. In the header of the grid, there is also a plus sign available for adding more columns to your grid. Since we only have five properties available, we don't have to add any more columns. If you want to add additional columns, you can do that by clicking on the + sign (see the plus sign on the far right of *Figure 13.10*).

At the top of the page, in the left corner, we see a green **New** button. Let's click on the **New** button and a sliding pane should open on the right. Here, we can find our form. With the form, we can not only create new records but also edit them. Because we are still in **BUILDERMODE**, all of the fields in our form are disabled (see *Figure 13.11*):

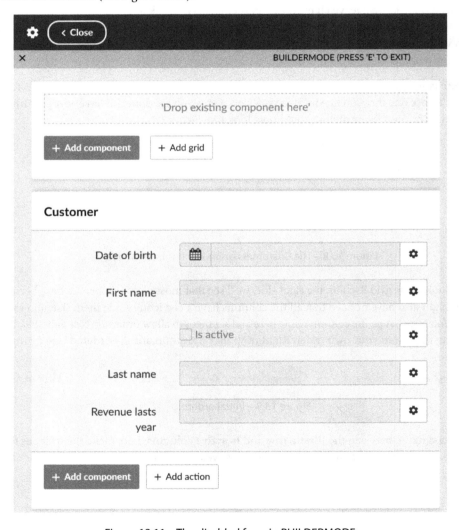

Figure 13.11 – The disabled form in BUILDERMODE

This also allows us to reorder our property fields in any order that we would like. Hover over one of the fields in your form. You should see an arrow shaped like a cross on the left side of the field. This arrow allows you to drag your field around. Grab the **First name** field and drag it to the top of the form, just above the **Date of birth** property. When the area above the **First name** property turns blue, release the mouse button and it should drop into that location (see *Figure 13.12*):

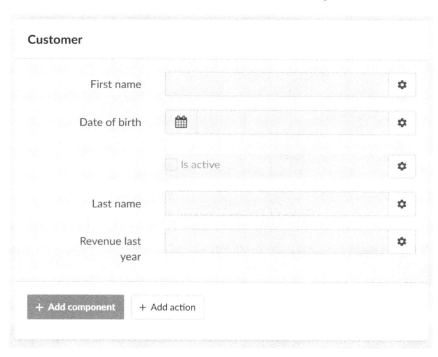

Figure 13.12 – The First name property placed on top

Let's order all these properties in a logical order:

1. **First name**
2. **Last name**
3. **Date of birth**
4. **Revenue last year**
5. **Is active**

Now, we are all set to enter our first data into our customer model. Exit Builder Mode by clicking on the wrench slider in the top-right corner or pressing the *E* button on your keyboard. Once you've done that, your form should look like *Figure 13.13*:

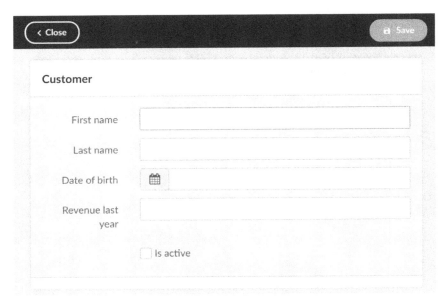

Figure 13.13 – Your updated form

If you select the **First name** field and press *Enter*, you should see three error messages appear. The fields next to which there are error messages are mandatory, meaning we can't save the form without filling those fields with information. Fill in all the fields in your form and hit the **Save** button to create your first record. After this, you should see your first record, as shown in *Figure 13.14*:

First name	Last name	Date of birth	Revenue last year	Is active
John	Doe	25-10-1987	250,000.00	✓

Figure 13.14 – Your first record in the grid view

Finally, we have some data in our data model. As you can see, it's actually quite easy to add a grid/ view from a model. All the options you set on your model are taking effect on your Back Office as well, so you don't have to worry about this anymore. Besides, after putting things in the right order, you're good to enter your first data. Of course, there is much more to do and learn, but I hope this first step, which is adding a new view to your back office, felt easy. As you can also see in *Figure 13.15*, at the top of the record you can see **1**:

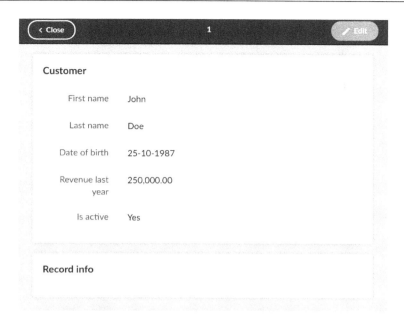

Figure 13.15 – Your first record in your form

This is the ID of the record in the database; it's automatically assigned and if you open up **Record info** by clicking on it, it should show you the **Created at** and **Updated at** properties. These properties are handled automatically as well, so you don't have to worry about them.

Remember that we also had an address model that had a relationship with the customer. Let's dive into that next. Because it's not visible right now, we would like to add one or more addresses to our records as well.

Adding a subview to your view

First, we'll discuss what a **subview** is and what it is used for in this section. Then, we'll see how we can add subviews and what kind of options they have. We'll also set up a view and then add data into a subview.

What is a subview?

What are subviews exactly? A subview is a view on top of your primary view that has a relationship with another model. This can be any kind of relationship. We have already added the **Customer** model as a view to our Back Office. So, in this case, we'll consider that to be our primary model. Our **Customer** model has a relationship with **Addresses**. We want to be able to see and add addresses to our customers. This is where the subview comes in.

Adding a new subview

You can add subviews in the view of your customer when you activate Builder Mode (with the wrench icon or the *E* key) and open one of your customer records. On the right side of the slider that opens, a button called **Add subview** should appear. Click on this button and a dialog should open up, as shown in *Figure 13.16*:

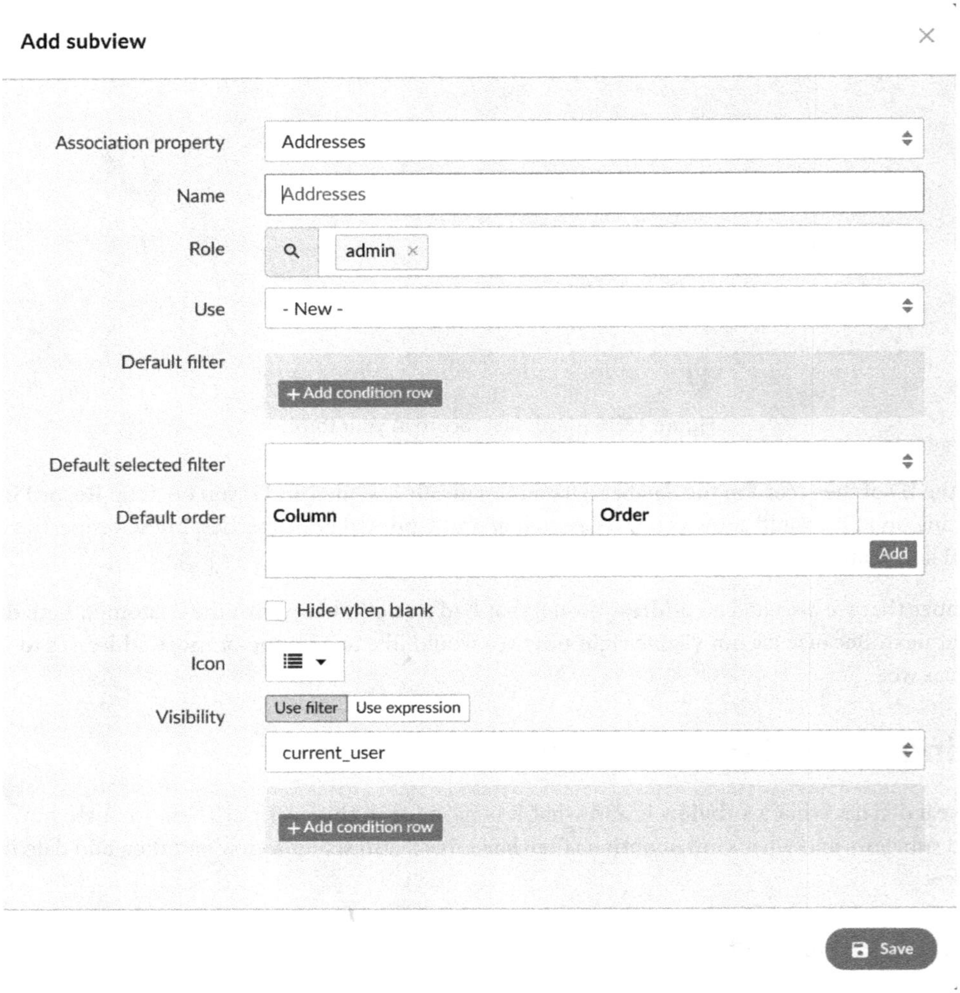

Add subview   ×

Association property	Addresses
Name	Addresses
Role	🔍 admin ×
Use	- New -
Default filter	+ Add condition row
Default selected filter	
Default order	Column Order Add
	☐ Hide when blank
Icon	☰ ▼
Visibility	Use filter Use expression
	current_user
	+ Add condition row

🖫 Save

Figure 13.16 – The subview dialog

As you probably noticed, this dialog has a lot of similarities with the dialog from when we created the view for the customer. That's because they are basically the same, except that this one is to create a subview and shows relational properties only.

As we can see, it already shows **Addresses** in the first field. It's sorted alphabetically, which is why **Addresses** is selected; if you click on the dropdown, you'll also see the orders there. In the **Name** field, we can see **Addresses**; again, it has been *plurified* for us. Of course, you can change it to anything you like. In the **Role** field, you can set up the roles that have access to this subview. Each view or subview can have specific access based on its role. Next is the filter; with this filter, you can prefilter the data before it's shown to the user. How can you use filtering in the platform? That's something we'll dive into in the next chapter about the Back Office. You can also set the ordering of your data by selecting the column that you want to be the default ordered column. Next is the icon selector; for a subview, you can also set the icon as you would for a normal view. The new field here is **Visibility**. This is another type of filter; it's of the permission type, kind of like **Role**. In this case, you can show the subview based on data from either your user or the customer.

Add the subview, based on the address, the standard admin role, and your own chosen icon. Don't change anything else right now. Let's close Builder Mode. Your screen should look like *Figure 13.17*:

Figure 13.17 – The Customer view with the Addresses subview

After you click on the **Save** button, you should see a subview called **Addresses** on the right side of your **Customer** form.

Setting up the view

The **0** adjacent to **Addresses** means that there are currently no records in **Addresses** that have a relationship with the customer. Let's add our first address to our customer by clicking on the plus sign adjacent to **Addresses**. As you will notice, two more sliding panes will open, one grid view for the addresses and the form for the new address. This is because we want to add a new address, so it is showing us the grid. Another thing you might notice is that all the form fields are in alphabetical order (see *Figure 13.18*). This isn't very logical for an address, so let's change that by turning on Builder Mode and putting it in the following order:

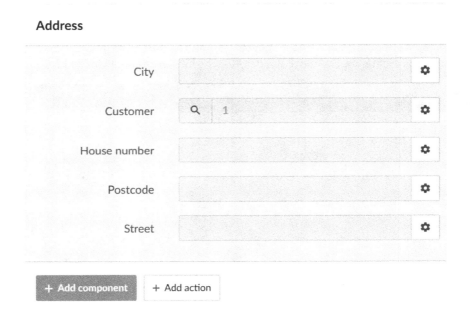

Figure 13.18 – Address in alphabetical order

As you can see, these are the fields from the preceding screenshot, in logical order:

1. **Street**
2. **House number**
3. **Postcode**
4. **City**

That leaves us with the **Customer** field—what is this field? It's the field that connects the address to the customer. Right now, it has a number in it, which represents the ID that the customer has been given by the platform (you can also change this in the data model by changing the label option there). Because we created the address directly from a customer, it automatically connects the customer to the address by filling in the ID. But we don't need to see this here. Of course, it could help you if you want to connect the address to another customer maybe, but that's not the use case here. So, let's remove the **Customer** field.

We can get rid of the **Customer** field in two ways. The easiest way is to drag it to the left. It will give us a popup asking whether we want to delete it. Another option is to click on the cog beside the **Customer** field. This will open a dialog showing us a lot of different options, which we'll go over in a later chapter. There is also a **Delete** button at the bottom of the dialog. If you click on the **Delete** button, it should also remove the **Customer** field. After you've removed that field and turned off Builder Mode, your **Address** subview should look like *Figure 13.19*:

Address

Street

House number

Postcode

City

Figure 13.19 – The Address form after removing the Customer field

Adding data to the subview

Now that we have the subview ready for use, let's add an address. Fill all the fields with an address of your choice. After saving the address, let's go back to our **Customer** view. You should see that it shows **1** now, beside **Addresses**. This means it has one address attached to it. You can add a few more addresses and you'll see that it keeps adding them nicely, while also allowing you to see all the addresses connected to this customer:

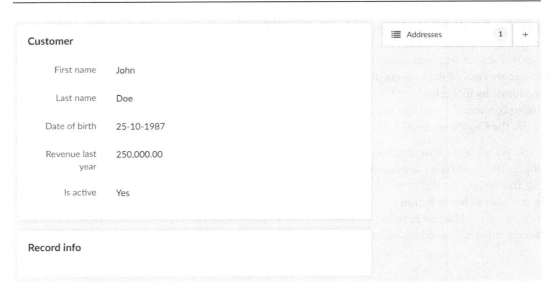

Figure 13.20 – Customer with one address

Click once more on the + sign beside **Addresses** and add another address. Just make an address up so we'll have two addresses. If you go to the grid view of **Addresses** (see *Figure 13.21*), you'll see that the order of the columns doesn't make sense here, either. Let's change it so it makes more sense. Enable Builder Mode to change the order. You can change the order in any way that you like; also you can make the **House number** and **Postcode** columns a little less wide so they take up less space and it's easier to see all the data:

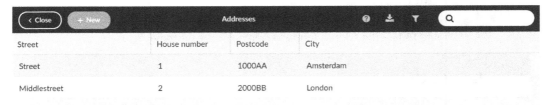

Figure 13.21 – The Addresses grid view with two records

Adding an extra property

Now that we have two addresses, we'll notice that one thing is missing. There is no real distinction between types of addresses, so we wouldn't really know what each address is for.

To solve this, we'll need to add a new property that can identify exactly which type of address it is. There are two ways of doing this. We can go back to the data model and go to the address model and add a new property, but there is also a quicker way of doing this. In Builder Mode, click on the + sign at the end of one of your columns. This + sign allows you to add new properties to your columns, but of course, we don't have this property yet.

In the new dialog that opens, there is also a shortcut for adding new properties to our model, as shown in *Figure 13.22*. Using the plus sign highlighted with a red border, we can quickly add a new property to our address model:

Figure 13.22 – The grid view with the plus sign highlighted

Once you've clicked on the plus sign on the right side of the dialog, a screen should appear that allows you to add a new property to your model:

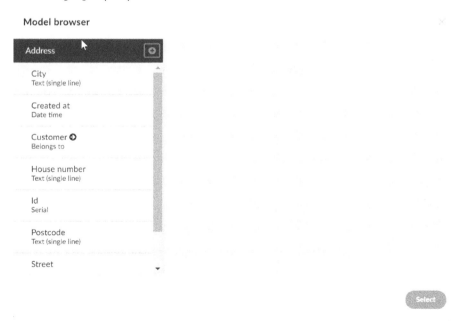

Figure 13.23 – Add property to column dialog with the add property (+) icon highlighted

Let's select the **List** type here. With this property, we can create a list of options for the user to choose from. In this case, we want our users to be able to choose what kind of address they are going to input into the record. Let's name this property Address type.

It should automatically fill in the label for us. As you might have noticed, this looks a little different from the previous screen that we've used for creating a new property. As mentioned before, this is the old interface for the platform, but it works basically the same as before. Now, we need to add some values to our list that our users can choose from.

Click on the **Add** button in the values, and let's add Home and Work here. This will allow users to choose between an address being identified as a home or work address. For now, to keep it simple, we'll just add these two values. Let's scroll down a little and also check the **Validate presence** option. This will make the **List** property mandatory to fill in for the users. That's all we need for this new property, so let's click on the **Save** button. Now, on the left side of the dialog, in the list of our properties, we should see the **Address type** property appear (see *Figure 13.24*). Let's select it and click on the **Select** button to add it to our grid view columns. Let's put it at the beginning of our columns.

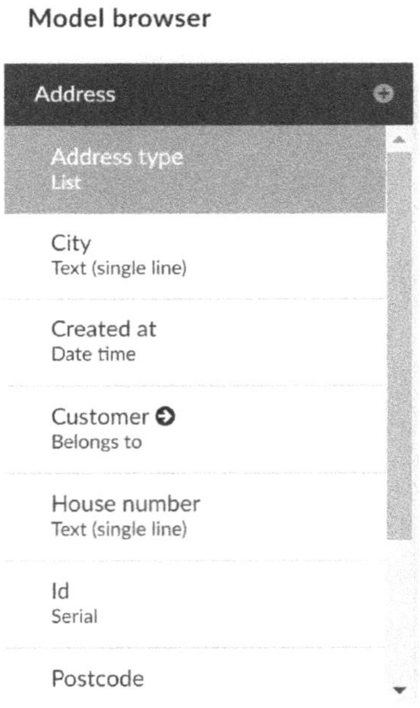

Model browser

Figure 13.24 – The Address type option added as a property

The column will appear empty at this point because it's a new property for all of our records and nothing has been set yet. So, we need to set **Address type** on all of our addresses as well. Since we only have two right now, we can do this manually. Let's open up one of our records. You'll notice that the address type is not yet visible in the form for our address, but luckily that is an easy fix. If you still have Builder Mode on, click on the **Add component** button in the **Address** section (see *Figure 13.25*):

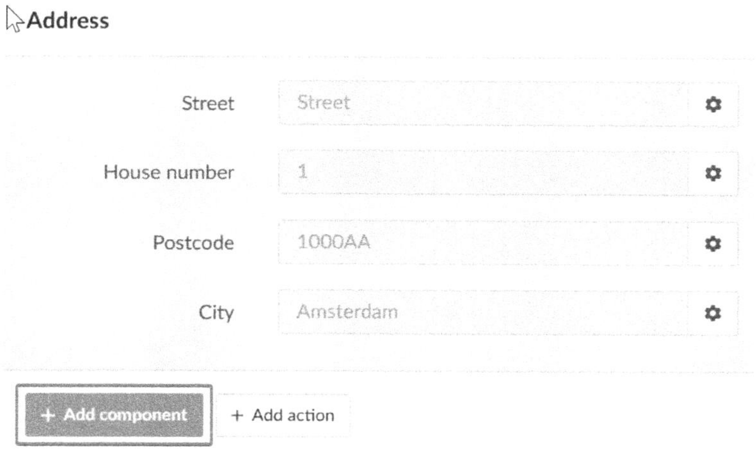

Figure 13.25 – The Add component button in the Address section

In the dialog that opens up after clicking on the **Add component** button, select **Address type** and press the **Select** button. This should add the **Address type** property to your form. Let's put the **Address type** property at the top of our form as well, so it's the first thing someone needs to select.

Now, you have two models inserted into the Back Office, using which you can insert data, edit the data, and view the data. Without writing a line of code, it's just the first step, because there is a lot more that you can do, but we're taking it one step at a time. Feel free to add the two other models, **Order** and **Order line**, to the Back Office now as well. You'll get an even better idea of how it works from this.

Summary

In this chapter, we have learned how to turn our models into Back Office views. This allows you to quickly turn your models into screens where you can add, edit, view, and delete the records in your models.

Also, you've learned how you can use relationships to create a subview between your models. This allows you to quickly add information for your customers, such as an address or even multiple addresses. Also, you've seen that you can quickly add a new property to your model and view without leaving the Back Office. This makes making quick changes very easy.

Index

Packt.com

Subscribe to our online digital library for full access to over 7,000 books and videos, as well as industry leading tools to help you plan your personal development and advance your career. For more information, please visit our website.

Why subscribe?

- Spend less time learning and more time coding with practical eBooks and Videos from over 4,000 industry professionals

- Improve your learning with Skill Plans built especially for you

- Get a free eBook or video every month

- Fully searchable for easy access to vital information

- Copy and paste, print, and bookmark content

Did you know that Packt offers eBook versions of every book published, with PDF and ePub files available? You can upgrade to the eBook version at packt.com and as a print book customer, you are entitled to a discount on the eBook copy. Get in touch with us at customercare@packtpub.com for more details.

At www.packt.com, you can also read a collection of free technical articles, sign up for a range of free newsletters, and receive exclusive discounts and offers on Packt books and eBooks.

Other Books You May Enjoy

If you enjoyed this book, you may be interested in these other books by Packt:

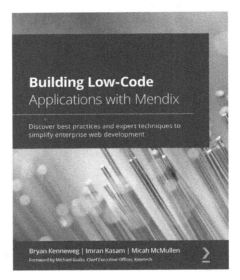

Building Low-Code Applications with Mendix

Bryan Kenneweg, Imran Kasam, Micah McMullen

ISBN: 978-1-80020-142-2

- Gain a clear understanding of what low-code development is and the factors driving its adoption
- Become familiar with the various features of Mendix for rapid application development
- Discover concrete use cases of Studio Pro
- Build a fully functioning web application that meets your business requirements
- Get to grips with Mendix fundamentals to prepare for the Mendix certification exam
- Understand the key concepts of app development such as data management, APIs, troubleshooting, and debugging

Democratizing Application Development with AppSheet

Koichi Tsuji, Suvrutt Gurjar, Takuya Miyai

ISBN: 978-1-80324-117-3

- Discover how the AppSheet app is presented for app users

- Explore the different views you can use and how to format your data with colors and icons

- Understand AppSheet functions such as yes/no, text, math, list, date and time and build expressions with those functions

- Explore different actions such as data change, app navigation, external communication, and CSV import/export

- Add/delete and define editing permissions and learn to broadcast notifications and inform users of changes

- Build a bot through the AppSheet Automation feature to automate various business workflows

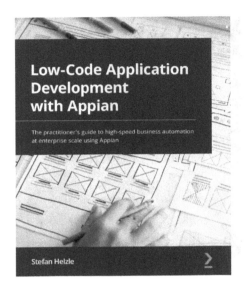

Low-Code Application Development with Appian

Stefan Helzle

ISBN: 978-1-80020-562-8

- Use Appian Quick Apps to solve the most urgent business challenges
- Leverage Appian's low-code functionalities to enable faster digital innovation in your organization
- Model business data, Appian records, and processes
- Perform UX discovery and UI building in Appian
- Connect to other systems with Appian Integrations and Web APIs
- Work with Appian expressions, data querying, and constants

Packt is searching for authors like you

If you're interested in becoming an author for Packt, please visit `authors.packtpub.com` and apply today. We have worked with thousands of developers and tech professionals, just like you, to help them share their insight with the global tech community. You can make a general application, apply for a specific hot topic that we are recruiting an author for, or submit your own idea.

Share your thoughts

Now you've finished *Democratizing Application Development with Betty Blocks*, we'd love to hear your thoughts! Scan the QR code below to go straight to the Amazon review page for this book and share your feedback or leave a review on the site that you purchased it from.

`https://packt.link/r/1803230991`

Your review is important to us and the tech community and will help us make sure we're delivering excellent quality content.

Download a free PDF copy of this book

Thanks for purchasing this book!

Do you like to read on the go but are unable to carry your print books everywhere? Is your eBook purchase not compatible with the device of your choice?

Don't worry, now with every Packt book you get a DRM-free PDF version of that book at no cost.

Read anywhere, any place, on any device. Search, copy, and paste code from your favorite technical books directly into your application.

The perks don't stop there, you can get exclusive access to discounts, newsletters, and great free content in your inbox daily

Follow these simple steps to get the benefits:

1. Scan the QR code or visit the link below

https://packt.link/free-ebook/9781803230993

2. Submit your proof of purchase
3. That's it! We'll send your free PDF and other benefits to your email directly

www.ingramcontent.com/pod-product-compliance
Lightning Source LLC
Chambersburg PA
CBHW060519060326
40690CB00017B/3322